T0135647

Air Entrainment and Macro Instability in Stirred Vessels

Der Technischen Fakultät der
Universität Erlangen-Nürnberg

zur Erlangung des Grades

DOKTOR – INGENIEUR

vorgelegt von

Jian Yu

Erlangen - 2003

Bibliografische Information Der Deutschen Bibliothek
Die Deutsche Bibliothek verzeichnet diese Publikation in der Deutschen
Nationalbibliografie; detaillierte bibliografische Daten sind im Internet über
http://dnb.ddb.de abrufbar.

ISBN 3-8325-0324-2

Logos Verlag Berlin
Comeniushof, Gubener Str. 47,
10243 Berlin
Tel.: +49 030 42 85 10 90
Fax: +49 030 42 85 10 92
INTERNET: http://www.logos-verlag.de

Gedruckt mit Unterstützung des

Deutschen Akademischen Austauschdienstes

Als Dissertation genehmigt von

der Technischen Fakultät der

Universität Erlangen-Nürnberg

Tag der Einreichung:	**24. Juni 2002**
Tag der Promotion:	**17. Januar 2003**
Dekan:	**Prof. Dr. rer. nat. A. Winnacker**
Berichterstatter:	**Prof. Dr. Dr. h.c. F. Durst**
	Prof. Dr.-Ing. K.-E. Wirth

to Nannan and Dalong

献给我的楠楠和龙儿，

父母以及所有关心我的亲人

FOREWORD

This thesis was accomplished during my study at the Institute of Fluid Mechanics (LSTM-Erlangen) as a DAAD (German Academic Exchange Service) Scholarship holder since May 1998.

First I would like to thank my mentor, Prof. Dr. Dr. h.c. F. Durst, for his guidance, including his leading role concerning the scientific methodology, continuous support and enthusiastic encouragement during my research work, without which the thesis could have never been finished.

My supervisor, Dr.-Ing. B. Genenger, supplied wonderful working conditions and atmosphere throughout my working time in the group. For his practical supervision and constant readiness for discussion, I am very grateful.

Prof. Dr.-Ing. K.-E. Wirth accepted the "Koreferat" not only of my thesis but also the annual report for DAAD, for which I am very grateful.

I would also like to express my appreciation to my friend and colleague Dr.-Ing. Schäfer for his comprehensive introduction to the necessary measuring techniques and fruitful discussions at the beginning of my research work in the group. I would like to thank my colleagues, Dr.-Ing. Wächter, Dr. Zivkovic, Dr. Sijercic and Mr. Nassar, for numerous helpful discussions and the detailed assistance.

For the financial support and friendly help of DAAD during my stay in Germany, I would like to express my sincere appreciation to all members of DAAD. Special thanks go to Ms. Schädlich and Ms. Böhme for their advice, assistance and friendship.

The research work in the later part of the thesis was supported by the European Union (EU) in its project "OPTIMUM", for which I wish to acknowledge. I would like to express my gratitude to Prof. M. Yianneskis of ECLAT at King's College London for fruitful discussions.

Furthermore, I am very grateful to the staff of LSTM-Erlangen. The happy cooperation with them in this work and also in daily life will remain permanently in my memory.

I take this opportunity to express my most sincere acknowledgement to my father-in-law, Prof. Dr.-Phys. Jun-Xiu Lin, for his meticulous guidance both in scientific promotion and in life philosophy.

Finally, I am very indebted to my family, above all my wife Nan and my son Runjia, for their continuous support and patience during the time I spent on my PhD work.

ABSTRACT

In open stirred vessels, the rotational speed of an impeller can only be increased within the limitation of gas bubble entrainment over the free surface, despite the fact that in most cases a higher mixing efficiency or other objectives can only be achieved by increasing the impeller speed up to a certain level. Numerous investigations in this field since 1958, which led to the establishment of empirical correlations between the critical rotational speed of impellers and the geometry as well as operating conditions have met their limitations in wider utilization in practice owing to the large deviations attributed to the lack of a fundamental understanding of the complicated mechanisms of air entrainment in stirred vessels.

In the present work, the mechanisms of surface aeration in stirred flows were determined by visual observations, numerical analysis and detailed flow field investigations. It was determined by visual observations that surface aeration is the direct common consequence of the formation of surface vortices and the disturbance on the surface vortices from the mean stirred flow. The surface vortices are the representation of PVC (precessing vortex core) motions in rotational flows, which results from different types of instabilities in rotational flows. This was confirmed by intensive numerical and experimental investigations on a simplified stirred flow, namely rotating disk flow in a square tank. Based on the mechanisms of surface aeration, a physical model applying dimensional analysis was introduced and verified in the present work, which is helpful during the design and optimisation of the stirring system with regard to either avoiding or enhancing surface aeration.

The PVC motion in stirred flows is the direct origin of the so-called MI (macro instability) phenomenon in stirred vessels which has attracted increasing attention of the research workers in the field of stirring techniques. The flow fields induced by two different types of impellers in stirred vessels were quantitatively investigated with the modern fluid mechanical measuring technique LDA (laser Doppler anemometry). The fundamental features of MI, the geometry influences on MI and the magnitude of MI in stirred flows were systematically studied. In addition, the turbulent wall jet flow before baffles contributing to bubble entrainment as disturbances was characterized, taking into account the MI influences. Finally, as improvements from the point of view of avoidance of air entrainment, shaft baffles and surface screens were introduced, and their validity was verified.

The present work has for the first time associated surface aeration in stirred flows with the precessing vortex motion and the macro instabilities. The mechanisms of surface aeration and MI in stirred vessels were revealed and further characterized.

TABLE OF CONTENTS

FOREWORD VII

ABSTRACT VIII

TABLE OF CONTENTS IX

NOMENCLATURE XIII

1   INTRODUCTION 1

   1.1   GENERAL CONSIDERATIONS 1
   1.2   SHORT LITERATURE SURVEY 3
      1.2.1   Mechanism of surface aeration 4
      1.2.2   Correlation models for onset of surface aeration in stirred tanks 6
      1.2.3   Macro instabilities in stirred vessels 9
   1.3   LAYOUT OF THE THESIS 11

2   THEORETICAL BACKGROUND 13

   2.1   ROTATIONAL FLOWS 13
      2.1.1   Governing equations 13
      2.1.2   Dimensional analysis and physical similarity 15
      2.1.3   Vortex classification 17
      2.1.4   Secondary flow around rotating bodies 18
      2.1.5   Swirl flow and vortex breakdown 19
      2.1.6   Rotational flow with free surface 20
      2.1.7   Instabilities in rotating disk flows 21
         2.1.7.1   Hydrodynamic instabilities 21
         2.1.7.2   Classification of instabilities in rotating disk flows 22
      2.1.8   PVC in swirl flows 28
   2.2   TURBULENCE AND STIRRING 30
      2.2.1   Turbulent flows and turbulence modelling 30
      2.2.2   Flow pattern in stirred vessels 32
      2.2.3   Characteristics of stirred flows 33
      2.2.4   Turbulence distribution in stirred vessels 35
   2.3   AIR ENTRAINMENT 36
      2.3.1   Introduction 36
      2.3.2   Air entrainment mechanisms 36

3   EXPERIMENTAL AND NUMERICAL TECHNIQUES EMPLOYED 39

   3.1   LASER LIGHT SHEET VISUALIZATION AND VIDEO ANALYSIS TECHNIQUE 39
   3.2   LASER DOPPLER ANEMOMETRY 40
      3.2.1   Basic principle of LDA 40
      3.2.2   LDA system employed 41
         3.2.2.1   Experimental set-up 41
         3.2.2.2   Measurement section 41
         3.2.2.3   LDA measuring system and data acquisition 44

    3.2.2.4 Sample size and error analysis 47

  3.3 NUMERIC METHOD AND FLOW COMPUTATION 48

   3.3.1 *General introduction*..*48*

   3.3.2 *Finite volume method*...*49*

    3.3.2.1 Discretization in space 49

    3.3.2.2 Discretization in time 50

   3.3.3 *Grid generation*...*51*

4 **INTRODUCTORY STUDIES OF ROTATING DISK FLOWS** **53**

  4.1 GEOMETRY SET-UP 53

  4.2 ANALYTICAL INVESTIGATION OF ROTATING DISK FLOW 54

   4.2.1 *Introduction* *54*

   4.2.2 *Analytical solution* *54*

  4.3 NUMERICAL INVESTIGATION OF ROTATING DISK FLOW....................................... 57

   4.3.1 *Introduction* *57*

   4.3.2 *Grid structure and level*..*58*

   4.3.3 *Results and discussion* *58*

    4.3.3.1 Steady flow calculations 58

    4.3.3.2 Unsteady flow calculations 64

  4.4 EXPERIMENTAL INVESTIGATION OF ROTATING DISK FLOWS 67

   4.4.1 *Laser sheet visualization in rotating disk flows in a square tank* *67*

   4.4.2 *Steady flow investigations in stirred vessels* *69*

    4.4.2.1 Introduction and experimental set-up 69

    4.4.2.2 Results and discussion.. 69

5 **PRIMARY RESULTS OF VISUAL OBSERVATIONS ON SURFACE AERATION** **73**

  5.1 GENERAL MECHANISMS OF SURFACE AERATION IN STIRRED VESSELS 73

   5.1.1 *Unbaffled vessels* *73*

   5.1.2 *Standard baffled vessels*...*75*

    5.1.2.1 Experimental set-up.. 75

    5.1.2.2 General mechanisms of surface aeration 75

    5.1.2.3 Other mechanisms of surface aeration.. 78

   5.1.3 *Positions of appearance of surface vortices in baffled vessels* *79*

    5.1.3.1 Radial stirrers 79

    5.1.3.2 Axial stirrers 81

   5.1.4 *Visualization of the surface vortex structure* *82*

    5.1.4.1 Introduction and geometry 82

    5.1.4.2 Results and discussions 83

  5.2 CRITICAL ROTATIONAL SPEED FOR THE ONSET OF SURFACE AERATION AND CORRELATIONS 85

   5.2.1 *Definition of different critical rotational speed* *85*

   5.2.2 *Comparison with the existing correlations* *86*

   5.2.3 *Theoretical analysis of the correlations*...*88*

6 **EVALUATION OF STUDIES ON SURFACE AERATION AND MACRO INSTABILITY** **93**

  6.1 FLOW PATTERN INVESTIGATION 93

   6.1.1 *Flow configuration* *93*

   6.1.2 *Unbaffled and non-fully baffled stirred flows* *95*

   6.1.3 *Fully-baffled stirred flow* *96*

    6.1.3.1 RT 96

    6.1.3.2 PBT 102

6.2   INVESTIGATIONS OF MACRO INSTABILITIES 104
  6.2.1   Times series analysis of velocities 105
  6.2.2   Spectrum analysis of macro instabilities 107
    6.2.2.1   Lomb periodogram algorithm 107
    6.2.2.2   MI dominant frequency spatial distribution 109
    6.2.2.3   Linearity MI dominant frequency 111
    6.2.2.4   Effect of sampling rate on MI 115
    6.2.2.5   Effect of the sampling time on MI 116
  6.2.3   Magnitude of MI 119
    6.2.3.1   Introduction 119
    6.2.3.2   Decomposition methods 120
    6.2.3.3   Processing of moving window averaging 121
    6.2.3.4   Intensity determination 123
    6.2.3.5   Spatial distribution of the magnitude of the low variations 125
    6.2.3.6   Dependency of the MI magnitude on rotational speeds 129
  6.2.4   Investigations of variations in Macro instabilities 131
    6.2.4.1   Introduction 131
    6.2.4.2   Results and discussion 131
6.3   WALL JET FLOW BEFORE BAFFLES 141
  6.3.1   Introduction 141
  6.3.2   Results of mean velocity profiles 142
  6.3.3   Results of semi-instantaneous velocity profiles 145
6.4   IMPROVEMENT OF AVOIDING SURFACE AERATION 147
  6.4.1   Shaft baffles 147
  6.4.2   Surface screens 149

7   CONCLUSIONS 153

8   REFERENCES.. 157

APPENDICES 171

  APPENDIX A: RUNGE-KUTTA METHOD WITH NEWTON-RAPHSON SEARCHING METHOD............................... 171
  APPENDIX B: SOLUTIONS OF ODES WITH FV METHOD APPLYING TDMA 176
  APPENDIX C: LOMB SPECTRAL ALGORITHM FOR UNEVENLY SAMPLED LDA DATA 182

SUMMARY IN GERMAN 185

INHALTSVERZEICHNIS 187

1   EINLEITUNG 191

  1.1   ALLGEMEINE EINFÜHRUNG.. 191
  1.3   INHALT DER DISSERTATION 194
7   ZUSAMMENFASSUNG 197

NOMENCLATURE

A	[-]	Experimental coefficient for correlations
$b_{1/2}$	[m]	Half with of the jet
B	[-]	Experimental coefficient for correlations
C	[m]	Impeller clearance from tank bottom
C_1, C_2, C_3	[-]	Experimental coefficient for correlations
D	[m]	Chamber diameter
D	[m]	Screen grid width
D	[m]	Impeller diameter
d_B	[m]	Mean bubble diameter
D_{disc}	[m]	Diameter of the disc (Rushton turbine)
D_{hub}	[m]	Hub diameter
D_{Shaft}	[m]	Diameter of the shaft
f_c	[Hz]	Mean frequency of primary circulation
f_D	[Hz]	Doppler frequency
F_j	[N]	Force
f_{MI}	[Hz]	Marco instability frequency
f_v	[Hz]	Vortex rotating frequency
g	[m/s²]	Gravitational acceleration
h	[J/kgK]	Enthalpy
H	[m]	Liquid height
H_b	[m]	Effective baffle height
h_{Hub}	[m]	Height of the hub
i, j, k	[-]	Index counter
k	[J/kg]	Turbulent kinetic energy
k	[-]	Wave number
l	[m]	Characteristic length
l	[m]	Grid spacing of the screen
L	[-]	Aspect ratio of the two-disk flow geometry
L_{Blade}	[m]	Length of the blade

m	[-]	Number of control volumes
n	[-]	Refractive index
N	[min^{-1}]	Rotational speed
N	[-]	Number of measurements
n_b	[-]	Number of baffles
$N_{1.in}$, N_{CSA2}	[min^{-1}]	Critical rotational speed of the first visible transferred bubble
N_{bubble}, N_{CSA1}	[min^{-1}]	Critical rotational speed of the first single visible bubble
N_{cloudy}	[min^{-1}]	Critical rotational speed of cloudy bubble cluster
N_{CSA}	[min^{-1}]	Critical rotational speed of surface aeration
N_{steady}	[min^{-1}]	Critical rotational speed of the steady surface aeration state
N_{vortex}	[min^{-1}]	Cirtical rotational speed of the first visible surface vortex
P	[Pa]	Pressure
P	[W]	Power consumption
P_0	[Pa]	Atmospheric pressure
Q	[m³/s]	Pumping volume flow rate, flow rate
Q_C	[m³/s]	Circulation volume flow rate
r, R	[m]	Radial coordinate, radius
r, z	[m]	Radial and axial cylinder coordinate
S_{Baffle}	[m]	Distance of the shaft baffle
t, T	[s]	Time
T	[m]	Vessel diameter
T	[°C]	Temperature
T_{Blade}	[m]	Thickness of the blade
u	[m/s]	Characteristic velocity
$u_\perp$	[m/s]	Perpendicular velocity component in the control volume
u_m	[m/s]	Local maximum velocity in the wall jet
U_i, U_j, U_k	[m/s]	Velocity
U_B	[m/s]	Bubble terminal rise velocity
$\bar{U}$	[m/s]	Average velocity

U, V, W, u, v, w	[m/s]	Radial, tangential and axial velocity components
u', v', w'	[m/s]	Radial, tangential and axial velocity fluctuations
V	[m³]	Vessel volume
V_{tip}	[m/s]	Tip velocity of the impeller
W	[m]	Width of the impeller blade
W_b	[m]	Width of the baffle
W_{Blade}	[m]	Width of the blade
W_{Screen}	[m]	Width of the surface screen
x_i, x, y, z	[m]	Cartesian coordinate
z	[-]	Confidence interval

Greek letters

β	[-]	Entrainment coefficient
δ	[m]	Boundary layer thickness
δ_{ij}	[-]	Kronecker delta function
Σ	[m²]	Cross-sectional area of vortex cylindrical chamber
μ	[Pas]	Dynamic viscosity
σ	[N/m²]	Surface tension
κ	[W/sm³]	Coefficient of heat conduction
τ_{ij}	[N/m²]	Molecular momentum transport
φ	[°]	Half cross angle of two laser beams
φ_0	[°]	Inclination angle of the streamline
Φ	[N/m²s]	Dissipation term
Φ	[-]	Transported quantity
Γ_Φ	[-]	Diffusion coefficient
S_Φ	[-]	General source term
ω, Ω	[rad/s]	Angular velocity
ε	[m²/s³]	Energy dissipation rate
ε	[°]	Spiral rolls wave front angle

ε_U	[-]	Errors for mean velocity
ε_{RMS}	[-]	Errors of RMS values
ϕ	[°]	Tangential component in cylindrical coordinates
ϕ	[°]	Angle from the middle plane of the impeller blade
λ	[m]	Wave length of the laser beam
ν	[m²/s]	Kinematic viscosity
θ	[°]	Angle before baffles
θ_c	[s]	Circulating time
Δt	[s]	Time step
Δx	[m]	Fringe spacing of the interference model
Δx	[m]	Cell spacing
ρ	[kg/m³]	Density

Dimensionless numbers

Bi	Baffle intensity number
Co	Courant number
fl	Flow number
Fr	Froude number
Mo	Morton number
Po	Power number
Re	Reynolds number
S	Swirl number
Sh	Strouhal number
Tu	Turbulence grade
We	Weber number
We'	Modified Weber number

Subscripts

B	Bubble

BP	Blade passing
c	Characteristic variables
CSA	Critical surface aeration state
Ft	Fully turbulent
G	Gas
L	Liquid
MI	Macro instability
nb, E, W, N, S, T, B	Neighbour CVs, correspondingly, east, west, north, south, top, and bottom
P	Node of the CV
Rand	Random fluctuation
Sb	Self baffled status
St	Steeping flow status

Superscripts

*	Dimensionless variables

Abbreviations

2D	Two-dimensional
3D	Three-dimensional
BPF	Blade passage frequency
CDS	Central differencing scheme
CNC	Computer numerical control
CV	Control volume
DFLDA	Diode fibre laser Doppler anemometry
DNS	Direct numerical simulation
FASTEST	Flow Analysis by Solving Transport Equations Simulating Turbulence
FV	Finite volume method
LDA	Laser Doppler anemometry

LES	Large-eddy simulation
MI	Macro instability
PBT	Pichted blade turbine
PC	Personal computer
PDA	Phase Doppler anemometry
PIV	Particle image velocimetry
POD	Proper orhtogonal decomposition
PVC	Precessing vortex core
RMS	Root-mean-square value
rev	Revolution
rpm	Revolutions per minute
RT	Rushton turbine
UDS	Upwind differencing scheme

1 INTRODUCTION

1.1 General Considerations

Stirred vessels are widely employed as operation units in the chemical, food and pharmaceutical industries, bio- and environmental technologies, and various other fields. Despite a hierarchical set of extremely different types of stirring systems in various application fields, the objectives of stirring processes can be classified into five general purposes according to Smith (1999) [117] and Zlokarnik (1999) [152]:

- Blending, homogenizing miscible liquids with or without chemical reactions
- Suspending solid particles
- Dispersing a gas through a liquid in the form of small bubbles
- Dispersing a second liquid, immiscible with the first, i.e. emulsification
- Promoting heat transfer

In blending and emulsion, only a single liquid phase takes part in processing, whereas in the other process objectives, gas or solid phases are normally involved. One stirring system must accomplish at least one objective. In practice, however, industrial processes often require more objectives having different levels of importance at the same time. The process objectives determine the equipment selection, design and operating conditions. Owing to the different levels of complexity of the process objectives and different orientated strategies, the optimisation or selection of equipment appropriate for all objectives involved in a process is often impossible [125]. An appropriate optimisation and design strategy oriented to the most important objectives relies on detailed flow information in stirred vessels.

Because of the above-mentioned importance of stirred vessel flow, numerous fluid flow investigations have been carried out, providing detailed information on flow properties. These investigations were strongly supported by recent developments in LDA, PDA and PIV techniques, providing means of experimental studies not available in the past [26, 97-111, 120, 144-146, 98, 99, 133, 147, 87]. Furthermore, numerical prediction procedures have been introduced into flow research and they have also been applied to stirred vessel flows. Through numerical simulations of stirred vessel flows, major insights into essential flow phenomena were obtained, providing a deeper understanding of agitated flows in stirred vessels [4, 6, 24, 114, 133-136]. The mean flow characteristics in single-phase flow in stirred vessels have been well investigated by different research groups employing both experimental and numerical methods, and enormous improvements in processes during stirring have been achieved [144]. However, many unsteady instantaneous mixing phenomena remain unclear in both experimental and numerical work, such as air entrainment over the free surface, the low frequency phenomenon [12], the precessing vortex phenomenon [117], etc.

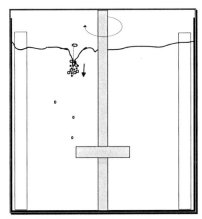

Figure 1.1: Surface aeration in stirred vessels

It is well known that air in opened or gas in closed vessels can entrain into the liquids over the free surface if the rotational speed of impeller increases to a certain level. Figure 1.1 depicts this phenomenon schematically. This phenomenon is also called surface aeration after Calderbank (1958) [17], and it often occurs under a stirrer rotational speed which is necessary to reach an efficient mixing level or to satisfy certain criteria such as solid particle suspension, etc. Therefore, surface aeration can affect the quality of the mixing process significantly through the operating conditions.

The lack of reliable information on the phenomenon of surface aeration motivated the present work. It started with a detailed literature survey which is summarized in the next section. The present work was important for an industrial project for the coatings industry carried out at LSTM-Erlangen [50], which was concerned with the avoidance of surface air entrainment. In this project it was shown that in spite of strong research requirements on this subject, few comprehensive investigations on the mechanisms of air entrainment in stirred vessels from fluid mechanics could be found. The first article associated with surface aeration appeared in 1958 [17]. However, later research work on this subject was limited to simplified dimensional analysis and setting-up of empirical correlations based on visual observations. Numerous correlations have been made between the critical rotational speed of the impeller and some substantial geometrical configurations and operating conditions including the impeller types, the impeller diameter, impeller submergence, the physical properties of the involved phases, etc. They suffered from major deviations owing to the very different types of stirring systems, due to the scarcity of a physical background, and thus offered little guidance in the design and optimisation phase in practice. Therefore, there are serious requirements for understanding the phenomenon of surface aeration in stirred vessels. From the standpoint of the avoidance of this phenomenon, the aim of the present work was to develop a guidance for selecting and designing a stirring system which can suppress surface aeration and at the same time still provide efficient stirring. Since there are already numerous correlations to predict the onset of surface aeration, the focus was also put on the judgement of the existing correlations in the literature and modifying them based on a better physical basis.

The author is convinced that the results obtained will be of great interest beyond his own project at LSTM-Erlangen, since surface aeration has also both positive and negative consequences, and for both consequences the present work would be helpful in guiding optimisation and design. An attractive feature of air entrainment or surface aeration is that it can bring or enhance gas-liquid contact by entraining bubbles from the open surroundings or vessel head. Thus gas dispersion can be achieved even without any additional sparger system. This type of aerator is called as surface aerator according to Forrester *et al.* (1997) [46]. Surface aerators are commonly used in biological waste water treatment equipment [150]. A greater advantage lies especially in many reactions such as hydrogenation, chlorination, oxidation, etc., where the conversion of gas per pass is small. In such instances it is highly desirable to recycle the unreacted gas from the headspace back to the reactors, since the gas may be highly toxic and its release could lead to environmental problems, or may be very expensive, making it worth recycling it. Gas-inducing performance characteristics are highly important and impeller design should be correspondingly orientated for such processes. More details were summarized by Forrester *et al.* [46].

On the other hand, air entrainment over the free surface can be undesirable. The entrainment phenomenon changes qualitatively the original behaviour of stirred media in the vessels. A typical example is stirring in the coating industry. The slightest bubble entrainment during mixing could lead to a significant impairment of the coating process, therefore, the quality of the coating process could not be ensured. An additional special post-procedure has to be introduced to get rid of bubbles from the coating fluid which in most cases is very difficult because of the high viscosity of the coating liquids. With massive bubble entrainment, the presence of air bubbles in the impeller region would cause a significant reduction in power consumption of the impeller, whereas in some processes such as solid suspension or heat transfer, sufficient power consumption is necessary. Also, the cavity generated by massive bubbles behind the impeller blades could cause mechanical problems such as shaft vibration, etc.

1.2 Short Literature Survey

Numerous investigations on mixing in ungassed liquids have mentioned the existence of surface aeration as a consequence of an increase in impeller speed. However, only a few studies have concentrated on the reason for this phenomenon and its onset conditions in the form of different types of empirical correlations. In this section, the results in the literature are summarized in two directions: the mechanisms of surface aeration and onset correlations. In the last decade, a large-scale long-time variation of the flow pattern in stirred tanks, the so-called "macro-instability (MI)", has attracted intensive attention from many research groups. It ill be shown that surface aeration can be associated with this. A short literature survey on this aspect is also given in this section.

1.2.1   Mechanism of surface aeration

When a liquid is stirred in an unbaffled vessel, a central vortex is formed around the stirrer shaft due to the centrifugal force of the impeller throwing the fluid out to the walls. The liquid in this vortex exhibits a solid body rotation, where the power consumption is low and very little mixing takes place. The depth of the vortex increases with increasing impeller speed. As soon as this vortex reaches the impeller, bubble entrapment occurs from the vortex causing surface aeration. This phenomenon was, as mentioned by Zlokarnik [152], even employed as an alternative type of aeration in the industry in earlier times. However, in most cases such aeration is undesirable owing to the partial absence of a liquid bearing, which leads to strong additional mechanical damage to the stirrer shaft. This vortex surface aeration in unbaffled vessels can be suppressed by eccentric mounting of or inclining of the stirrer [125].

In baffled tanks, the baffles act as obstacles to solid body rotation. The flow rebounds from the baffles, causing a high degree of turbulence. Power consumption increases and the vessel contents are well mixed. The liquid level in such baffled tanks is more or less wavy and live. Air entrapment can occur at such liquid surfaces. Clark and Vermeulen (1964) [25] proposed that a continuous increase in the impeller to tank volume ratio, i.e. D^2W/T^2H, should bring the system into a region where the tank baffling is overcome, and vortex and circulatory motion is again approached. This regime could be characterized by a sudden drop in power number Po as the Reynolds number Re is increased owing to the accumulation of bubbles behind the impeller (cavitation), whereas such a sudden drop does not take place in a closed vessel. In addition, they further observed the liquid surface under the conditions of surface aeration in vessels agitated by a four-bladed flat-paddle impeller in baffled tanks, and drew the conclusion that, on the surface, a region of high shear exists. This region occurs owing to a combination of two flows in opposite directions: first the impeller discharge flow, and second the flow that rebounds from the baffles. The surface does not remain stationary, but appears to be in random oscillatory motion. Along the upper seam made by this surface, air is introduced into the liquid.

Greaves and Kobbacy (1981) [52] described surface aeration in stirred ungassed vessels in great detail. Their observations were carried out in a fully baffled flat-bottomed tank with a diameter of 0.203 m. Three sizes of Rushton turbines with different D/T ratios were used. It was observed that small vortices are formed owing to the flow pattern at the free surface. They emphasized that surface aeration starts at a low impeller speed prior to any drop in Po. Small eddies are formed at random in a region between the shaft and the vessel walls, then combine into a stronger eddy to form a hollow vortex. Gas bubbles are drawn in at the bottom of such eddies and carried down to the impeller region. As the impeller speed increases, the eddies residing at the surface grow in intensity. More gas is drawn into the liquid, producing a corresponding increase in gas hold-up near to the surface. At still higher impeller speed, also larger sizes and numbers of bubbles are transported to the impeller region, which leads to an increase in the flow rate of gas to the impeller. Veljkovic *et al.* (1991) [131] also mentioned that at the free surface eccentric and unsteady vortices exist.

Albal *et al.* (1983) [1] reported that, for surface aeration under unsparged conditions, depending on the agitation speed, there are three distinct regimes of mass transfer from the gas headspace to the liquid. At very low stirring speed, the mass transfer occurs mainly by diffusion through the gas-liquid interface. This regime was called "surface diffusion". As the stirring speed increases the convective forces and the rate of the surface renewal at the interface determine the mass transfer. This regime was called "surface convection". In this regime, Davies and Lozano (1979) [29] pointed out that the Prandtl sized eddies, and even larger eddies, determine mass transfer rates at a free surface. At still higher stirring speeds, the surface vortices entrain a large amount of gas in the form of gas bubbles which become distributed throughout the vessel. This regime was called "surface entrainment" and is normally responsible for massive surface aeration, whereas at high stirring speeds the first two mechanisms are negligible.

Many other authors have also described the mechanism of surface aeration. Brauer and Schmidt-Traub (1973) [10] pointed out that at very high impeller speed, the baffles introduce a strong second circulatory flow at the surface, leading to the formation of many small vortices. These vortices are place and time dependent and have a short lifetime. They stressed that the formation of such vortex aeration is not preventable. Takase *et al.* (1983) [122] considered the small waves at the free surface in a stirred tank and concluded that the formation of small waves at the free surface is responsible for the surface aeration. Liepe *et al.* (1998) [73] proposed that under insufficiently baffled conditions the bubble entrapment occurs owing to the formation of a central vortex near the shaft, and with an increase in the number or/and width of the baffles, a fully baffled state can be reached where *Po* does not increase further with the baffle intensity numbers. Under these conditions the bubbles can be entrained into the fluid mainly by the turbulence near the free surface.

Chandrasekhar (1961) [20] carried out instability analysis in unbounded systems (e.g. in the absence of vessel walls) for inviscid fluids. He started with the Navier-Stokes equations for the gas and the liquid phases. Any disturbance of the gas-liquid interface was assumed to be in sinusoidal form with a characteristic frequency. By using the linear stability theory, the relative velocity between the gas and liquid phases at which the gas-liquid interface becomes unstable was calculated. This instability of the gas-liquid interface is called "Kelvin-Helmholtz" instability. For air-water system this value is close to 6.6 m/s. For bounded systems, this value is likely to be substantially lower, mainly because the presence of vessel walls and baffles generate turbulence which also results in disturbance of the gas-liquid interface.

According to the above description of the mechanism of surface aeration by different workers, surface aeration can be considered as a two-step process, the first step being the entrapment of gas at the liquid surface due to high degree of turbulence and the second the carriage of these bubbles from the liquid surface into the bulk. The ability of the impeller to generate sufficient turbulence to entrap the gas bubbles and the ability to generate a favourable liquid-phase flow pattern for efficient dispersion of the entrapped bubbles are the important parameters governing the phenomenon of surface aeration.

1.2.2 Correlation models for onset of surface aeration in stirred tanks

Owing to the complexity of the flow pattern in stirred tanks and of the mechanism of the surface aeration, a direct theoretical analysis of the onset of the surface aeration is almost impossible. A number of studies were carried out to correlate the critical impeller speed for the onset of surface aeration with the geometry of the system and the operating conditions, in order to predict the critical impeller speed during the design of stirred tanks.

Over the years, various workers have defined and used different criteria to determine the critical speed. In unsparged vessels the critical rotational speed of surface aeration N_{CSA} can be determined visually as the speed at which gas bubbles start to become entrapped at the surface [17, 52, 57, 121, 121-124]. Tanaka *et al.* (1986) [124] defined two values of N_{CSA}: first, if the impeller speed is increased above a certain value where the bubble entrapment just begins, this being denoted the critical impeller speed for enfolding bubbles; second, if the impeller speed is increased further, entrapped bubbles are dispersed throughout the reactor, this speed being denoted the critical impeller speed for steady-state dispersion. The value of the critical impeller speed for steady-state dispersion was found to be larger than that for enfolding bubbles by about 20%.

Quantitatively, the critical speed is determined by two other levels of surface aeration. The first is the sudden drop of the power number Po, which can be exactly obtained in the power measurement. Several workers [17,52, 57] used such a critical impeller speed in their correlations. The other method was defined and applied by Calderbank *et al.* [18], Miller (1974) [33], Veljkovic *et al.* [131], etc. The critical impeller speed for surface aeration was taken as the value of the impeller speed at which the rate of mass transfer increases rapidly. Both criteria are only appropriate for massive air entrainment. From the point of view of the avoidance of surface aeration, visual observations were more often carried out.

Most of the studies on the critical speed for surface aeration have been empirical in nature. These studies determine N_{CSA} by any of the methods mentioned above and correlate it with the operating conditions such as the impeller submergence, the impeller bottom clearance, the geometric characteristics of the system like the type and diameter of the impeller, the physical properties of the liquid phase, etc. These studies and the correlations are listed in Table 1.1.

So far only a few studies have been directed at modelling of the phenomenon of the onset of surface aeration. Greaves and Kobbacy [52] assumed that the rate-controlling step in surface

Impeller Type	System Geometry	Correlation	Reference
Flat paddle	$T = 0.25\ m$	$Fr \cdot \dfrac{D^2 W}{T^2 H} \cdot (W/H)^{2/3} = 0.005$	Clark and Vermeulen (1964) [25]
RT	$T=0.19\ m$ $H=0.205\ m$	$\dfrac{N_{CSA}D}{U_B} = A + B\dfrac{T}{D}$	Boerma and Lankester (1968) [94]
RT	$T=0.165\text{-}2.6\ m$ $D/T=0.13\text{-}0.7$	$N_{CSA} = 1.55 \cdot + (\dfrac{T}{D})(\dfrac{H-C}{T})^{0.5}(\dfrac{\sigma g}{\rho_L})^{0.25}$	Dierendock et al. (1971) [94]
RT	$T=0.22\ m$ $D/TD=0.4\text{-}0.6$	$N_{CSA} = 2.1 \cdot D^{-0.5}$	Matsumura et al. (1977) [94]
RT	$D/T=0.3$ $V_L=16\text{-}70\ l$	$N_{SCA} = kd^{0.61}V^{*0.23}(\rho_{L\,Ref}/\rho_L)^{0.32}(V_{L\,Ref}/V_L)^{0.095}$	Sverak and Hruby (1981) [121]
RT	$T=0.2\ m$ $D/T=0.38\text{-}0.67$	$N_{CSA} = C\dfrac{(T^2 H^2)^{1/3}}{D^2}(1-\dfrac{C}{H})^{1/3}(P/P_0)^{-0.13}$	Greaves and Kobbacy(1981) [52]
RT	-	$\dfrac{N_{CSA}d^{1.98}}{D^{1.1}} = \dfrac{1.65}{Ne^{0.125}}(\eta_L/\eta_G)^{0.031}(W/D)^{0.625}$	Joshi et al. (1982) [65]
RT	$T=0.12\text{-}0.2\ m$ $D=0.05\text{-}0.1\ m$ $t/H=0.25\text{-}0.75$	$N_{CSA} = 126(\mu_L/\gamma)^{0.94}(D/T)^{-2.3}(H/T)^{0.44}[(H-C)/H]^{0.3}$	Tanaka et al. (1986) [124]
RT, PBT, propeller	$T=0.21\text{-}0.54\ m$ $D=0.4\text{-}0.13\ m$ $H/T=0.56\text{-}1.29$	$N_{CSA} = AT^f D^a C^b H^c$	Heywood et al. (1986) [59]
Vaned disc turbine,	$T=0.57\ m$ $D/T=0.3\text{-}0.5$	$N_{CSA} = K\dfrac{(T^2 H^2)^{1/3}}{D^2}(1-C/H)^{1/3}$	Ram Mohan and Kolte (1988)
RT	$T=0.2\text{-}0.675\ m$ $D/T=1/3$ $H/T=1$ $C/T=0.33$	$N_{CSA}D - 0.732 = 2812.1\mu_B D$	Veljkovic et al. (1991) [131]
RT, PBT	$T=0.15\text{-}0.64\ m$ $D/T=0.2\text{-}0.5$ $C/D=1/3\text{-}1$	$N_{CSA} = \sigma^\alpha v^\beta T^\omega \left(\dfrac{H-C}{H}\right)^\gamma \left(\dfrac{T}{D}\right)^\delta$	Ditl et al. (1997) [34]
RT	$T=0.1\text{-}1\ m$ $D/T=0.19\text{-}0.61$ $C/D=0.17\text{-}0.5$	$N_{CSA} = \dfrac{2Po^{1/3}}{\sqrt{c}}\dfrac{ND}{d_B}\dfrac{D}{T}\dfrac{1}{F_G}$	Zehner et al. (1999) [149]

Table 1.1: Summary of correlations developed by different research groups

aeration is the ability of the impeller to generate the liquid flow to carry the bubbles from the interface to the bulk of the vessel. They made such an assumption because, near the gas-liquid interface, very high gas hold-up was observed. Based on this, they arrived at a semi-empirical relationship for the critical impeller speed for the onset of surface aeration. The details of their derivation are given below. For a Rushton turbine, the average velocity of the liquid leaving the impeller in the radial direction U_r is given by

$$U_r = C_1 \frac{ND^2}{r}, \tag{1.1}$$

where N denotes the rotational speed, D the impeller diameter and r the radial distance. C_I is a constant to be determined experimentally. Since the average liquid velocity in the vessel is of the same order of magnitude, the average velocity $\bar{U}$ of the circulating liquid in the tank with diameter T and filling height H is given by

$$\bar{U} = 1.387 \frac{ND^2}{(T^2 H)^{1/3}}. \tag{1.2}$$

A gas bubble entrapped at the surface will be transported to the bulk if $\bar{U}$ and the terminal rise velocity of the bubble U_B satisfy $\bar{U} > U_B$ or $\bar{U} = C_2 U_B$. Substituting this in Equation (1.2), solving for N and denoting it N_{CSA} gives

$$N_{CSA} = C_3 \frac{(T^2 H^2)^{1/3}}{D^2} (1 - \frac{C}{H})^{1/3} (P/P_0)^{-0.13}, \tag{1.3}$$

where C is the impeller clearance and P and P_0 denote the working and atmospheric pressure, respectively. The values of C_3 were found to be, e.g., 0.48 and 0.43 for distilled water and tap water, respectively.

Joshi et al. (1982) [65] independently developed a similar approach. For RT, the liquid circulation velocity near the wall is given by the following equation:

$$U_r = 0.53(D/W)ND(D/T)^{7/6}, \tag{1.4}$$

where W denotes the blade width. The average circulation velocity in the vessel can be approximately taken as U_r. They further assumed that surface aeration will occur when the downward liquid circulation velocity equals the terminal rise velocity of the gas bubble. They assumed that the average bubble diameter d_B is given by

$$d_B = 1.21 \frac{\sigma^{0.6}}{(P/V)^{0.4}(\rho_L)^{0.2}} (\mu_L/\mu_G)^{0.1}, \tag{1.5}$$

where the subscripts G and L represent the gas and liquid phase, respectively, σ is the surface tension, P/V the averaged energy disspation rate, ρ the density and μ the viscosity. The rise velocity U_B is

$$U_{B\infty} = 0.71\sqrt{gd_B}. \tag{1.6}$$

Substituting Equation (1.5) in (1.6) and assuming that U_B is proportional to U_r as given in Equation (1.4), we obtain

$$\frac{N_{CSA}D^{1.98}}{T^{1.1}} = \frac{1.65}{Po^{0.125}}(\mu_L/\mu_G)^{0.031}(W/D)^{0.625}\,, \tag{1.7}$$

where Po is the power number. For an air-water system and RT with $W/D = 0.2$, and $Po = 5$, the above equation reduces to

$$\frac{N_{CSA}D^{1.98}}{T^{1.1}} = 0.34\,. \tag{1.8}$$

Since there were substantial variations in the experimental conditions used by different workers, there are also large variations in the correlations proposed. For example, Veljkovic *et al.* [131] proposed that $N_{CSA} \propto 1/D$, whereas Greaves and Kobbacy [52] reported that $N_{CSA} \propto 1/D^2$. For a clearer comparison, a series of Rushton turbines varying in D/T ratio in a 400 mm diameter tank were selected. The predicted surface aeration critical rotational speed was obtained with the correlations listed in Table 1.1 and the comparative results are plotted in Figure 1.2. It is obvious that considerable deviations exist between different correlations, limiting the effective prediction of the onset of the surface aeration in practical applications.

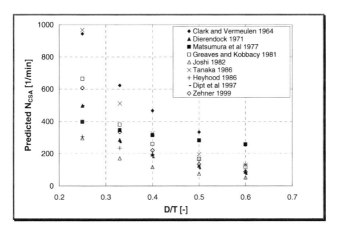

Figure 1.2: Comparison of the predicted values of N_{CSA} with different correlations

1.2.3   Macro instabilities in stirred vessels

In the last 10 years, increasing attention has been devoted to the large-scale long-time flow variations in stirred vessels. This phenomenon was often termed macro instability (MI). This macro instability indicates the existence of eddy fields whose time and length scales exceed considerably those associated with blade passages and small-scale turbulence. This large-scale low-frequency phenomenon was observed by Rao and Brodkey (1972) [101] for a paddle impeller, appearing in the form of a lower band of dominant frequencies in the one-dimensional velocity spectrum. Since 1988, macro instabilities for different types of impellers have been regularly reported in the literature, e.g. by Winardi et al. (1988, 1991) [140, 141], Haam and Brodkey (1992) [56], Kresta and Wood (1993) [68], Bruha et al. (1994-1996) [14, 15, 16], Chapple and Kresta (1996) [29], Myers et al. (1997) [87], Montes et al. (1998) [86], Roussinova et al. (2000) [103, 104], Hasal et al. (2000) [58], etc.

Winardi et al. [141] were among the first to report that the instantaneous circulation pattern for a paddle impeller is different from the mean velocity field as measured using LDA. In a second paper, Winardi et al. [140] applied several flow visualization techniques and LDA to examine the time-varying flow pattern of a marine propeller. They found that the flow pattern was asymmetric with respect to the impeller shaft, and the flow pattern changes were random in order and the lifetime of a given pattern could range from half a second to several minutes. Kresta and Wood [68] observed variations in the bulk circulation pattern when they analysed the flow field of a PBT using flow visualization, LDA and spectral analysis. A later paper by Chapple and Kresta [23] considered the influence of geometric parameters such as off-bottom clearance, impeller diameter and number of baffles on flow stability for two axial impellers. It was shown that the geometric variables have the most significant influence on the macro instability.

Bruha et al. [14-16] investigated four- and six-bladed 45° PBT impellers. They studied the effect of impeller rotational speed and the different geometries in which macro instabilities occur. Their investigations covered two impeller diameters ($D/T = 1/3$ and 1/4) and three off-bottom clearances ($C/T = 0.33$, 0.4 and 0.5). They observed a vortex which appears as a welling up to the fluid surface, indicating the appearance of the MI. To capture the shape and dimensions of the large vortex they used a video camera. A round probe above the fluid surface was used to measure the frequency of this "surface swelling". They concluded that this frequency f_{MI} is linearly dependent on the impeller rotational speed and therefore the dimensionless frequency normalized by the impeller speed N (f_{MI}/N) is a constant for the turbulent regime. The effect of geometry was not systematically examined, although several variations were studied. More recent work from this group used flow visualization and a special mechanical device, the so-called "tornadometer", which is placed below the surface in the vicinity of the impeller and deflects whenever MI are strong enough to change the direction of the mean flow and deflect the target. The authors were able to confirm their earlier report that the frequency of the MI is linearly related to the impeller rotational speed [15]. They also observed that the low-frequency MI are accompanied by changes in the angle of the impeller discharge flow and the appearance of an unstable secondary circulation loop, which Kresta and Wood [68] mentioned in their earlier work.

Montes et al. [86] used LDA measurements, flow visualization, spectral analysis and wavelet transforms to analyse MI in a stirred tank equipped with a six-blade 45° PBT impeller ($D/T = 1/3$, $C/T = 0.35$, four baffles). They confirmed that the occurrence of MI is accompanied by the presence of a large vortex in the upper part of the vessels, and it is linearly coupled with the frequency of the impeller revolution. The RMS velocity underwent a sudden rise in fluctuation intensity at $Re = 600$. This rise might correspond to the first appearance of the MI. For $Re > 600$, the frequency spectrum also show a distinct peak for low-frequency oscillations. This implies that both the transient and fully turbulent flow regimes are subject to this type of instability. Hasal *et al.* [58] implemented a new type of analysis of MI. Analysis of the velocity time series was done with the proper orthogonal decomposition (POD) technique. The contribution of the MI motion to the RMS velocities was estimated quantitatively. The authors showed that the MI has no coherent frequency for the geometry investigated. The frequency of the MI could not be determined in zones close to the vessel wall, or at high Re. Roussinova *et al.* [104] investigated the influence of geometric variations, and applied a velocity decomposition technique to analyse the magnitude of the MI.

From the above introduction to research on MI, it can be summarised that MI is an important component of the large-scale motions in a stirred tank. It has implications for many of the operations which are carried out in stirred tanks and for the structural integrity of the vessel components. The MI produces a broad, low-frequency band in the power spectrum. It does not necessarily show a coherent frequency, although this may be observed for some cases. The time scale of the MI is of the order of the tank turnover time. No frequencies are observed in the interval between the MI and the blade passage frequency, so these motions are clearly associated with the mean velocity determined over a short time interval, rather than with the turbulent fluctuations.

1.3   Layout of the Thesis

Since the stirred flow is, first of all, a rotational flow, the basic definitions and features are introduced in the next chapter. As a common feature of rotational flows, instabilities and the most advanced investigations on them for confined rotating disk flows are discussed in that chapter. As the consequence of instabilities, vortex breakdown and its precessing motion are introduced in Chapter 2. Turbulence and its characterization in stirred flows as well as its influence on air entrainment are also introduced in this chapter. The principles and the application of the experimental techniques employed in the present work are introduced in Chapter 3. In addition, the numerical methods applied in the present work including the basic background of the finite volume method and grid generation are also summarized Chapter 3.

Chapter 4 describes the introductory studies of stirred vessel flows. The flow induced by a rotating disk, which is assumed to have a more simplified flow pattern, allows one to penetrate into basic flow instabilities by employing analytical, numerical and experimental investigations. The precessing motion of the vortex core is observed in both experimental and numerical investigations. In Chapter 5, the primary observation results for stirred flows with air entrainment are described. The mechanism of surface aeration is studied by visual observations and verified by analytical and experimental investigations. It is concluded that the initial surface aeration can be well associated with the precessing of the vortex. Based on the results of visual observations, a physical model based on dimensional analysis is introduced and verified Chapter 5.

In Chapter 6, the related flow characterizations with regard to the formation of surface vortices are carried out, the flow pattern at the surface, the macro instabilities in stirred vessels as a consequence of the precessing vortex core, the magnitude of macro instabilities and the wall jet flow before baffles contributing to vortices, are analysed and evaluated. Two possibilities to avoidance of surface aeration in stirred vessels are introduced and also verified. Finally, general conclusions and recommendations for future work are given in Chapter 7.

2   THEORETICAL BACKGROUND

The flow in stirred vessels is well characterized as a three-dimensional, highly turbulent complicated flow [125, 152]. However, since motions in stirred vessels are generated by the rotation of the stirrer elements, stirred flows possess fundamental features and characteristics of rotational flows. Rotational flows have been studied extensively for a variety of reasons, and their technological applications vary from industrial fields, e.g. stirring, centrifugal pumps, turbo machinery, cyclone separators and so on, to natural phenomena such as tornadoes, hurricanes, ocean circulations, etc. Actually, most rotational flows are unsteady. Turbulent rotational flows like stirred flows form their circulation pattern and normally can be characterized by turbulence, micro-scale dissipation and macro-scale instabilities [79]. It is well known that an axially symmetrical flow will lose its stability and become non-axially symmetrical, regardless of the axis symmetry of the boundary conditions. The disturbances can include buoyancy, centrifugal forces, Coriolis forces, forces of gravity, viscous forces, etc.

In this chapter, the major features of rotational flows which are associated with the present work are introduced, including the governing equations, similarity analysis and classification of vortices. Instability investigations for the most confined rotational flow, rotating disk flow, are described with emphasis on hydrodynamic instabilities, vortex breakdown, swirl flow and precessing vortex core (PVC). Turbulence plays an essential role in mixing. The most important characteristics in stirred flows related to turbulence are introduced. Finally, the state and the fundamental mechanisms of air entrainment are considered.

2.1   Rotational Flows

2.1.1   Governing equations

The motion of fluids is governed by the conservation laws of mass, momentum and energy, and can be expressed in Cartesian coordinates as follows:

Mass conservation (continuity equation):

$$\frac{\partial \rho}{\partial t} + \frac{\partial}{\partial x_i}(\rho U_i) = 0, \tag{2.1}$$

where ρ is the density, t time and U_i the Cartesian velocity component in the x_i direction.

Momentum conservation (Navier-Stokes equations):

$$\rho\left[\frac{\partial U_j}{\partial t} + U_i\frac{\partial U_j}{\partial x_i}\right] = -\frac{\partial P}{\partial x_j} - \frac{\partial \tau_{ij}}{\partial x_i} + F_j, \tag{2.2}$$

where P denotes the pressure, τ_{ij} the stress tensor and F_j the body forces applied to the fluid.

For Newtonian media, the stress τ_{ij} can be expressed as [36]

$$\tau_{ij} = -\mu \left[\frac{\partial U_j}{\partial x_i} + \frac{\partial U_i}{\partial x_j} \right] + \frac{2}{3} \delta_{ij} \cdot \mu \cdot \frac{\partial U_k}{\partial x_k}, \tag{2.3}$$

where μ represents the dynamic viscosity and δ_{ij} the Kronecker delta function.

To close the equation system, the thermodynamic and transport properties have to be related by equations of state:

Energy conservation:

$$\rho \left[\frac{\partial h}{\partial t} + U_i \frac{\partial h}{\partial x_i} \right] = \frac{\partial P}{\partial t} + U_i \frac{\partial p}{\partial x_i} + \frac{\partial}{\partial x_i} \left(\kappa \frac{\partial T}{\partial x_i} \right) + \Phi, \tag{2.4}$$

where h denotes the enthalpy, κ the coefficient of heat conduction, T the absolute temperature and Φ the dissipation term.

For Newtonian fluids with ρ = constant and μ = constant, Equations (2.1) and (2.2) can be simplified considerably:

Continuity equation: $\quad\quad\quad\quad\quad\quad \dfrac{\partial U_i}{\partial x_i} = 0. \tag{2.5}$

Momentum equation: $\quad\quad \rho \left(\dfrac{\partial U_j}{\partial t} + U_i \dfrac{\partial U_j}{\partial x_i} \right) = -\dfrac{\partial P}{\partial x_i} + \mu \dfrac{\partial^2 U_j}{\partial x_i^2} + F_j. \tag{2.6}$

If the density is known, the above four equations form a closed equation system and can describe the flow field completely.

The governing equations can be transformed between different orthogonal coordinate systems depending on the corresponding flow configuration considered. Among them, the cylindrical coordinate system is widely applied in practice for rotating flows. Details on the transform of governing equations can be found in Durst (2000) [36]. The equations for fluids with ρ = constant and μ = constant can be rewritten in the following form:

Continuity equation: $\quad\quad\quad \dfrac{\partial U}{\partial r} + \dfrac{1}{r} \dfrac{\partial V}{\partial \phi} + \dfrac{\partial W}{\partial z} + \dfrac{U}{r} = 0, \tag{2.7}$

where U, V, W denote the velocity components in radial r, tangential ϕ and axial z direction, respectively.

Navier-Stokes equations for r, ϕ, z components, correspondingly:

$$\rho\left(\frac{\partial U}{\partial t}+U\frac{\partial U}{\partial r}+\frac{V}{r}\frac{\partial U}{\partial \phi}-\frac{V^2}{r}+W\frac{\partial U}{\partial z}\right)$$
$$=-\frac{\partial P}{\partial r}+\mu\left[\frac{\partial}{\partial r}\left(\frac{1}{r}\frac{\partial}{\partial r}(rU)\right)+\frac{1}{r^2}\frac{\partial^2 U}{\partial \phi^2}-\frac{2}{r^2}\frac{\partial V}{\partial \phi}+\frac{\partial^2 U}{\partial z^2}\right]+F_r, \tag{2.8}$$

$$\rho\left(\frac{\partial V}{\partial t}+U\frac{\partial V}{\partial r}+\frac{V}{r}\frac{\partial V}{\partial \phi}+\frac{UV}{r}+W\frac{\partial V}{\partial z}\right)$$
$$=-\frac{1}{r}\frac{\partial P}{\partial \phi}+\mu\left[\frac{\partial}{\partial r}\left(\frac{1}{r}\frac{\partial}{\partial r}(rV)\right)+\frac{1}{r^2}\frac{\partial^2 V}{\partial \phi^2}+\frac{2}{r^2}\frac{\partial V}{\partial \phi}+\frac{\partial^2 V}{\partial z^2}\right]+F_\phi, \tag{2.9}$$

$$\rho\left(\frac{\partial W}{\partial t}+U\frac{\partial W}{\partial r}+\frac{V}{r}\frac{\partial W}{\partial \phi}+W\frac{\partial W}{\partial z}\right)$$
$$=-\frac{\partial P}{\partial z}+\mu\left[\frac{\partial}{\partial r}\left(\frac{1}{r}\frac{\partial}{\partial r}(rW)\right)+\frac{1}{r^2}\frac{\partial^2 W}{\partial \phi^2}+\frac{\partial^2 W}{\partial z^2}\right]+F_z. \tag{2.10}$$

2.1.2 Dimensional analysis and physical similarity

Complete solutions to engineering problems can be seldom obtained by analytical methods alone and experiments are usually necessary to determine completely the way in which one variable depends on others. Dimensional analysis is widely applied in practice to connect the relevant variables and suggest the most effective way of grouping the variables. It enables the magnitudes of individual quantities relevant to a physical problem to be assembled into dimensionless number groups. Thus, in general, dimensional analysis results in a reduction in the number of parameters requiring separate consideration in an experimental investigation of a problem. A physical process behaves "similarly" in a model and its full-sized counterpart.

For example, let us take the governing equations for Newtonian fluids described in Section 2.1.1. The dimensionless form of partial differential equations can be expressed by dimensionless values which are normalized by characteristic values which are subscripted with c, e.g.

$$U_j=U_c\cdot U_j^*,\ t=t_c\cdot t^*,\ \rho=\rho_c\cdot\rho^*,\ P=\Delta P_c\cdot P^*,\ \tau_{ij}=\tau_c\cdot\tau^*,\ g_j=g_c\cdot g_j^*,\ \mu=\mu_c\cdot\mu^*, \tag{2.11}$$

where all values with superscript * are dimensionless. Substitution of the normalized values into continuity Equation (2.1) gives

$$\frac{\partial \rho}{\partial t}+\frac{\partial}{\partial x_i}\left(\rho U_i\right)=\frac{\rho_c}{t_c}\cdot\frac{\partial \rho^*}{\partial t^*}+\frac{\rho_c\cdot U_c}{L_c}\frac{\partial(\rho^*U_i^*)}{\partial x_i^*}=0.$$ (2.12)

This can be rewritten as

$$\underbrace{\frac{L_c}{t_c\cdot U_c}}_{Sh}\cdot\frac{\partial \rho^*}{\partial t^*}+\frac{\partial(\rho^*U_i^*)}{\partial x_i^*}=0,$$ (2.13)

where Sh denotes the Strouhal number. This equation indicates that similar solutions from the continuity equation can be obtained if the Strouhal numbers of the described flows are the same. In a similar way, for Navier-Stokes Equations (2.2) we obtain

$$\rho^*\left[\underbrace{\frac{L_c}{t_c\cdot U_c}}_{Sh}\frac{\partial U^*}{\partial t^*}+U_i^*\cdot\frac{\partial U_j^*}{\partial x_i^*}\right]=-\underbrace{\frac{\Delta P_c}{\rho_c\cdot U_c^2}}_{Eu}\cdot\frac{\partial p^*}{\partial x_j^*}-\underbrace{\frac{\tau_c}{\rho_c\cdot U_c^2}}_{1/Re}\cdot\frac{\partial \tau_{ij}^*}{\partial x_i^*}+\underbrace{\frac{g_c\cdot L_c}{U_c^2}}_{1/Fr}\cdot\rho^*\cdot g_j^*.$$ (2.14)

From Equations (2.14), three new dimensionless numbers come into being, the Euler number Eu describing the ratio of pressure forces to inertial forces, the Reynolds number Re and the Froude number Fr. Similarity solutions can be obtained if the dimensionless numbers are same.

However, mathematical solutions often do not exists for engineering problems in practice. Dimensional analysis is more often carried out with the help of the Π-theorem in which the dimensional homogeneity of physical process is considered. More details about dimensional analysis can be found in Zlokarnik (1991) [151]. Owing to the great complexity of stirred flows, dimensional analysis is very often applied in process analysis.

Physical similarity makes the results performed under different sets of conditions compatible. It includes geometry similarity (shape), kinematic similarity (motion) and dynamic similarity (forces). For the first two cases, the ratios of the lengths (more often termed scale factors) for geometry and the ratios of both lengths and time intervals for kinematic similarity have to be kept same, whereas for dynamic similarity, ratios of forces need to be similar. It is usually impossible to satisfy all of forces applying on a fluid simultaneously, therefore, normally only the most important forces are taken into account. The important ratios of forces widely used in stirred flows are listed in Table 2.1. Here it is valid that $\nu=\frac{\mu}{\rho}$.

Group	Symbol	Interpretation	Definition
Reynolds number	Re	Ratio of the inertial to viscous forces	$\dfrac{\rho l^3 \times (u^2/l)}{(\mu u/l) \times l^2} = \dfrac{lu}{\nu}$
Froude number	Fr	Ratio of the inertial to gravity forces	$\dfrac{\rho l^3 \times (u^2/l)}{\rho l^3 g} = \dfrac{u^2}{lg}$
Weber number	We	Ratio of the inertial to surface tension forces	$\dfrac{\rho l^3 \times (u^2/l)}{\sigma l} = \dfrac{u^2 \rho l}{\sigma}$

Table 2.1: Important ratios of forces in stirred flows

Here l and u denote the characteristic length and velocity in the flow, respectively, and g, ν and σ are the gravity acceleration, kinematic viscosity and surface tension of the fluid, respectively.

2.1.3   Vortex classification

Rotational flow is the rotating motion of a multitude of material particles [79], e.g. fluids around a common centre. This type of motion is normally termed vortex motion. The angular velocity of matter at a point in continuum space is called vorticity. The angular momentum of a particle is proportional to its velocity and its distance from the centre around which it is rotating, and the angular momentum of a closed system is a constant. If the distance is diminished, the velocity must increase correspondingly, so high angular velocities can occur without a high energy input. To preserve its angular momentum, a fluid particle in such a flow tends to accelerate its azimuthal velocity component as it moves towards to the rotating axis of the helical swirl flow, since the angular velocity is inversely proportional to the square of the particle trajectory radius. To generate a concentrated spiral vortex, only slight initial rota-

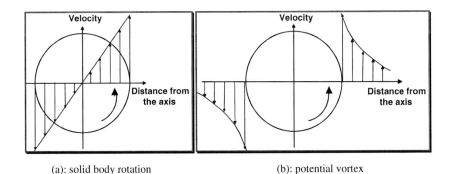

(a): solid body rotation   (b): potential vortex

Figure 2.1: Velocity distribution in two different types of vortices

tion in the fluid is needed and it is more or less self-induced.

The vortex motion occurring only in a circular plane is called a circular vortex. Circular vortices can be classified into two basic types: solid body rotation (also often called forced vortex) and the potential vortex (free vortex). In the first case, the velocity of points increases linearly with their distance from the centre as shown in Figure 2.1-a. The angular velocity is constant anywhere in the fluid, and in this case the fluids rotate like a solid body. Figure 2.1-b depicts the second rotation type. The fluid velocity is at its highest and equals the velocity of the rotating rod at the rod's surface owing to the adherence. With increasing distance from the rod the velocity diminishes in inverse proportion to the distance. This type of fluid motion is called a potential vortex, the fluid has no vorticity. In the first case no shear force occurs and thus no energy is needed to maintain its motion, whereas for the potential vortex, kinetic energy has to be provided constantly to make shearing possible between liquid sheets, and is ultimately transformed into heat inside the fluid ("dissipation"). Fluid motion composed of a potential vortex and a solid body rotation is called a "Rankie vortex". For a steady, circular motion without a velocity component normal to the rotation plane, the "Rankie vortex" is the only possible vortex whose velocity is zero both at the centre and far away from it.

2.1.4   Secondary flow around rotating bodies

The observation of the famous "tea-cup phenomenon", in which tea leaves collect in the centre of a cup when the liquid is stirred, indicated the existence of a secondary flow pattern in the rotational flow since ancient Greece times. The fluid dragged alone by a rotating body moves in a rotatory way. The simplest rotation is a potential vortex which can be obtained ideally by the rotation of an infinitely long circular cylinder in a fluid at rest. Actually, for cylinders of finite length or noncircular bodies the flow is more complex. When the body rotates very slowly, the surrounding fluid is dragged with it in concentric circles. At higher Reynolds number, a secondary flow is superimposed on the rotation as sketched in Figure 2.2-a for a cylinder and Figure 2.2-b for a rotating disk. Centrifugal forces push the fluid away from

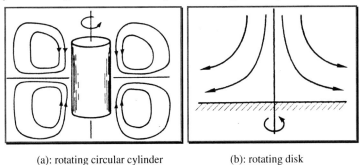

(a): rotating circular cylinder   (b): rotating disk

Figure 2.2: Schematic representation of the secondary flow around rotating bodies

from the centre and toward the edge. For mass conservation, the fluid is sucked into the vicinity of the rotating body. With increasing Reynolds number the flow starts to become unstable. For instance, Gregory *et al.* (1955) [54] studied the instability in the boundary layer of a rotating disk. Above $Re = \Omega R^2/\nu = 190,000$ the flow becomes unstable, and a vortex ring forms, which separates the laminar from the turbulent flow.

2.1.5 Swirl flow and vortex breakdown

Most vortices in nature, however, have a spatial structure, i.e. the pathlines are not perpendicular to the axis of rotation but have a component parallel to it. Vortices with an axial component are called swirling flows. In the presence of a radial velocity component, a spatial vortex tend to be a spiral vortex. A special and simple spatial vortex is the helical vortex, and it can be imagined that a constant axial motion is superimposed on the solid body rotation. The fluid travels along a helical path.

As for any viscous flow, the basic regime parameter of flow in a vortex chamber is the Reynolds number $Re = Q/(d\upsilon)$, where Q is the flow rate, d is the chamber diameter and υ is the kinematic viscosity. In addition, to characterize the degree of flow swirling in a vortex chamber, a swirl number is introduced. Various ways of determining this parameter exist. The simplest expressions represent the ratios of maximum tangential velocity to a maximum axial velocity or averaged tangential velocity to an averaged axial velocity. The most widespread definition is (Gupta *et al.* 1984 [55])

$$ S = \frac{F_\phi}{F_z d / 2}, \tag{2.15} $$

where $F_\phi = \int_\Sigma \rho VWr\mathrm{d}\Sigma$ is the axial component of the angular momentum flux, $F_\phi = \int_\Sigma (p + \rho W^2)\mathrm{d}\Sigma$ is the axial component of the momentum flux, V and W are the tangential and axial velocity components, r is the radial coordinate, ρ is the fluid density, d is the chamber diameter, p is the atmospheric pressure and Σ is the cross-sectional area. For turbulent swirl flows, corresponding time-averaged values are used.

Swirling flows in general, which do not receive rotational energy downstream, do not always decay by spreading vorticity in the way that the decaying potential vortex does. The tangential velocity decreases downstream. The flow is stable as long as the pressure also decreases in the downstream direction. However, the pressure has a tendency to increase because the tangential velocity decreases. If this tendency prevails at the axis, a separation point (stagnation point) may occur at the axis, which indicates zero axial flow. The vorticity tube abruptly expands and forms a closed egg-shaped region or a spiral. This phenomenon is called "vortex breakdown".

2.1.6   Rotational flow with free surface

The liquid surface in an open vessel will change
its horizontal form according to different types
of rotational flows. For example, in an open rotating cylinder, the filled fluid is caused to rotate
together. Owing to the centrifugal force, the free
surface tends to deviate to a parabolic shape
from the horizontal form. This is the well known
"bucket experiment" [79]. The reason for the
surface changing can be explained in the following analysis.

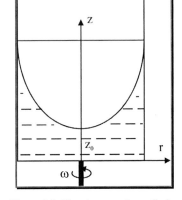

Figure 2.3: Flow in a rotating cylinder

Figure 2.3 displays the flow in a rotating cylinder under the conditions of solid body rotation. If
we choose the rotating cylinder coordinate system as the inertial reference system, namely the
coordinate system rotates with the same angular velocity as the fluid. After achieving a steady
state, there is no relative motion between fluid particles in the chosen coordinate system (U, V,
$W = 0$). From Equations (2.8), (2.9) and (2.10) we obtain

$$\frac{\partial P}{\partial r} = \rho g_r,; \quad \frac{1}{r}\frac{\partial P}{\partial \phi} = \rho g_\phi; \quad \frac{\partial P}{\partial z} = \rho g_z. \tag{2.16}$$

For a constant angular velocity ω, $g_r = \omega^2 r$, $g_\phi = 0$ and $g_z = -g$, we have

$$\frac{\partial P}{\partial r} = \rho \omega^2 r \quad \Rightarrow \quad P = \frac{1}{2}\rho \omega^2 r^2 + F_1(\phi, z), \tag{2.17}$$

$$\frac{1}{r}\frac{\partial P}{\partial \phi} = 0 \quad \Rightarrow \quad P = F_2(r, z), \tag{2.18}$$

$$\frac{\partial P}{\partial z} = -\rho g \quad \Rightarrow \quad P = -\rho g z + F_3(r, \phi), \tag{2.19}$$

where F_1, F_2 and F_3 are the assumed function of the corresponding coordinates.

From the above equations for constant C:

$$P = C + \frac{1}{2}\rho \omega^2 r^2 - \rho g z \qquad \text{for } 0 \le \phi \le 2\pi. \tag{2.20}$$

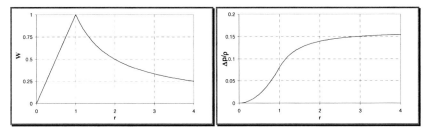

(a): Tangential velocity distribution   (b): Pressure distribution

Figure 2.4: Velocity and pressure in a Rankie vortex

Substituting the boundary conditions $P = P_0$ for $r = 0$, $z = z_0$, we obtain $C = P_0 + \rho g z_0$, where P_0 represents the atmospheric pressure. Thus, Equation (2.20) can be written as

$$P = P_0 + \frac{1}{2}\rho\omega^2 r^2 - \rho g(z - z_0) \qquad \text{for } 0 \leq \phi \leq 2\pi. \qquad (2.21)$$

The surface of the liquid where $P = P_0$, can be described by

$$z = z_0 + \frac{\omega^2}{2g}r^2 \qquad \text{for } 0 \leq \phi \leq 2\pi. \qquad (2.22)$$

For such a solid body rotation, the free surface of the liquid follows the parabolic pressure distribution. The pressure drop $P - P_0$ is proportional to the square of the rotational speed. If the rotational speed is high enough, the air dissolved in water near the axis of the rotation can reach the evaporation point at normal room temperature. This phenomenon is also called cavitation and thus can induce air entraining.

In an analogous way, the free liquid surface of a "Rankie vortex" can be acquired. Figure 2.4 represents the radial profile shape of the tangential velocity W and pressure drop Δp in a "Rankie vortex". The adjoining point of two basic vortex types is at $r = 1$.

2.1.7 Instabilities in rotating disk flows

2.1.7.1 Hydrodynamic instabilities

It is well known that an axisymmetric rotating flow will lose its stability and become non-axisymmetric, regardless of the axisymmetry of the geometry configurations. Instability is a mechanism by which a fluid accommodates to strong forces and new flow patterns are created. When a flow is disturbed and the disturbance decreases with time so that after a while the original state of flow is regained, the flow is said to be stable. In contrast, for unstable flows, instability leads to a new kind of flow. The interplay of the participating forces determines the occurrence of instability. These forces can be of several types: inertial forces, centrifugal and Coriolis forces, buoyancy, gravitation, electromagnetic forces, surface tension and geometry imperfection.

According to Gregory *et al.* [54], the hydrodynamic instability can generally be classified into three types:

- Dynamic instability, produced by centrifugal force or an external field of force due to gravity. Typical examples are Taylor vortex and Görtler vortex instability.

- Inflectional instability, which occurs when the two-dimensional velocity distributions contain a point of inflection at infinite Reynolds number. A physical explanation in terms of momentum transfer has related the instability to the nature of the vorticity field of the mean flow. The disturbances take the form of progressive waves with periodicity in the direction of the mean flow.

- Viscous instability, which can happen with many two-dimensional velocity distributions which are stable at infinite Reynolds number and are destabilised only by the action of viscosity. The disturbances are fed by the Reynolds shear stress transferring energy from the mean flow. It appears that the disturbance is dominated by viscosity in two regions in the flow: one is a secondary boundary layer adjacent to the wall, and the other is a viscous layer at the location where the velocity of the progressive wave disturbance equals the velocity of the fluid.

2.1.7.2   Classification of instabilities in rotating disk flows

Owing to the very complicated flow features of real applications and phenomena in nature, confined rotating flows have attracted much interest as a test case for contemporary ideas for a better understanding of the complicated real flow features and associated instability characteristics. A typical example of confined rotating flow is the flow due to a rotating disk. Since for infinite geometries exact similarity solutions of Navier-Stokes equations can be obtained for the stationary axisymmetric basic flow, these similarity solutions allow the reduction of Navier-Stokes equations to a system of ordinary differential equations, details of which will be further described in Section 4.2.

A considerable amount of research has been devoted to the stability of the rotating disk family. It can be classified into one- and two-disk flow problems, the latter including the case where only one disk rotates while the other is at rest. According to the limiting cases of angular velocity of the disk Ω and the angular velocity Ω_f of the rotating fluid, the basic flow boundary layer can be correspondingly referred to as the von Kármán layer ($\Omega \neq 0$, $\Omega_f = 0$), Bödewadt layer ($\Omega = 0$, $\Omega_f \neq 0$) and Ekman layer ($\Omega \approx \Omega_f$) (Faller 1963 [42]). The basic flow patterns of the first two cases were well summarized by Schlichting (1979) [112]. For the two-disk flow, two different solutions of the basic flow confined between two infinite parallel disks were derived by Batchelor and Stewartson, respectively. More details were reviewed by Zandbergen and Dijkstra (1987) [148]. The Batchelor solution showed that three flow regions develop at high rotation rates, having the structure of two shear layers near the walls bounding an inviscid core rotating at constant angular velocity $\Omega_c = \beta\Omega$, where β is the constant entrainment coefficient, nearly equal to 1/3. When the Reynolds number is decreased, the Batchelor stationary flow with separated boundary layers evolves towards a purely viscous flow with merged boundary layers. However, numerous experimental (e.g. Gauthier *et al.* 1999 [48]) and numerical (Dijkstra and Van Heijst, 1983 [33]) studies have shown that the angular velocity of the core deviates from the self-similar solution. The Stewartson solution was for smaller rotation rate. It was shown that the Batchelor flow similarity solution was the only stable one [27].

Linear stability analyses of the similarity solutions have been preformed by Faller (1991) [43] and Lingwood (1997) [75], for the instability and for the above-mentioned three layers, and demonstrated two instability types, normally referred as Type I and II by Faller and Kaylor (1966) [42] or Type B and A according to Greenspan (1968) [53]. Both instabilities appear in the form of regular systems of annular and spiral rolls confined to the disk boundary layer. These roll-like patterns differ in orientation, phase velocity and wavelength. Type I instability results from an inviscid mechanism due to unstable inflection points in the boundary layer velocity profile and also referred as "cross-flow" instability. Instability Type II relates to the combined effects of Coriolis and viscous force, and is dominant in low Reynolds number but stable in inviscid fluid flow. The spatial structure of both instabilities consists of travelling vortices in the boundary layers, and the orientation of their wave fronts with respect to the basic flow is positive for Type I, and negative for Type II. More recently, a third instability, Type III, and absolute instability induced by the coalescence of Types III and I in the form of convection of spirals in the Ekman layer were pointed out by Lingwood [75].

The instabilities of the axisymmetric flow of a two-disk flow field are governed by two parameters: the Reynolds number *Re* and the aspect ratio *L = H/R*, where *H* is the cavity height and *R* the radius of the cylinder. In the literature, the Reynolds number is based either on the external radius of the cavity, $Re_R = \Omega R^2 / v$, or on the height of the cavity, $Re = \Omega H^2 / v$, where v is the kinematic viscosity of the fluid. With different aspect ratios *L*, different types of axisymmetric or non-axisymmetric instabilities can be observed.

Experimental work on one-disk family

The most important experimental work for
the one-disk family was that of Gregory *et
al.* [54] in the von Kármán layer. Using a
china-clay visualization technique, they
were able to show the "cross-flow" instability (Type I). The boundary layer flow of
a rotating disk can be separated into three
regimes. Within the inner radius the flow is
purely laminar, and beyond the outer radius,

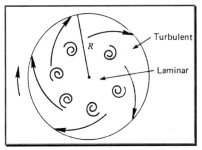

Figure 2.5: Instability Type I

wholly turbulent. In between, a series of traces in the form of equi-angular closely spaced
spirals indicate the presence of stationary vortices lying stationary relative to the disk surface.
The start point of *Re* for instability is about 190,000 and for transition 280,000. Figure 2.5
depicts these flow regimes. The co-author Stuart in the same paper [54] showed analytically
that type I instability results from an inviscid mechanism due to unstable inflection points in
boundary layer velocity profiles.

Low aspect ratio of the two-disk flow family

Different patterns resulting from instabilities appear in the flow between a rotating and a stationary disk enclosed by a stationary sidewall. According to the different ranges of the aspect
ratio applied, the studies can be classified into two groups: small aspect ratio ($L\square$ 1) and
large aspect ratio ($L \geq 1$). Totally different instability forms arise in different ranges of aspect
ratios.

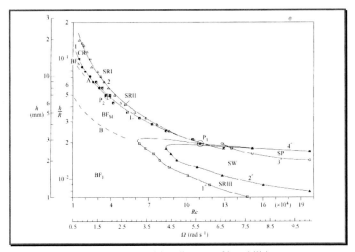

Figure 2.6: Transition diagram of instabilities

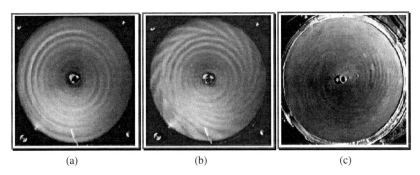

(a)   (b)   (c)

Figure 2.7: (a): Circular rolls (CR), (b): co-existence of CR and spiral rolls (SRI) and (c): spiral
rolls (SRII). Basic flow direction is clockwise.

For small aspect ratios, the most recent experimental work of Schouveiler *et al.* (2001) [113] summarized the different types of flow and instabilities resulting in a transition diagram as depicted in Figure 2.6. Three types of basic flows are distinguished by curves A and B and are denoted BF_J (adjoined layers), BF_S (separated layers) and BF_M (mixed state between the first two). At high aspect ratios in the transition diagram, curves 1 and 2 are the thresholds for the circular rolls (CR, see Figure 2.7-a) and the spiral rolls (SRI see Figure 2.7-b), respectively. The spiral rolls can be characterized by the angle of the wave front ε (for CR, $\varepsilon = 0$). For SRI, this value is about 25°. The frequency associated with the CR instability is strongly associated with the rotational frequency N of the rotating disk and varies in the range from $3N$ to N radially inwards, while the frequency of spiral rolls is not locked to the disk rotation frequency N. Spiral rolls break the axisymmetry of the flow emerging from the peripheral region of the flow with 16-24 arms depending on the initial conditions.

For intermediate aspect ratio ($1.79 \times 10^{-2} \leq L \leq 7.14 \times 10^{-2}$), only one mode of instability appears corresponding to curve $1''$, and also has a spiral form (SRII) as shown in Figure 2.7-c. However, it differs from the SRI in spiral angle (12-15°), these spirals are stationary in the reference of the laboratory. Other instability types arise as the aspect ratio decreases. More

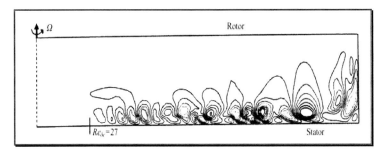

Figure 2.8: Time-dependent axisymmetric instability of the Bödewadt layer, represented by
the instantaneous iso-lines of the fluctuation of axial velocity

details can be found in Schouveiler *et al.* [113].

The annular and spiral rolls were also observed in numerical simulations. In recent work by Serre *et al.* (2001) [116], the flow instabilities in geometries similar to that of Schouweiler *et al.* [113] (L = 0.2 or 0.5) were analysed using three-dimensional direct numerical simulation (DNS). The transition in both the Ekman and Bödewadt layers was investigated accurately. The instability features in the Bödewadt layers have a lower threshold than those in the Ekman layers. Axisymmetric instabilities appear in the form of circular rolls

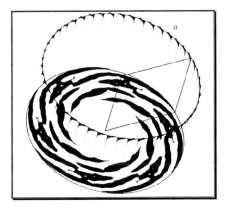

Figure 2.9: Pure spiral travelling mode, demonstrated by the iso-axisymmetric instability of the Bödewadt layer

and develop from stationary circular vortices into oscillatory travelling vortices as shown in Figure 2.8. By superimposing a three-dimensional perturbation on the axisymmetric basic flow, similar pure spiral rolls ($\varepsilon = 25.7°$) arise as depicted in Figure 2.9. Spiral rolls co-exist with circular rolls. The instabilities are three-dimensional and non-axisymmetric. Unlike in experiments where the onset of the three-dimensional flows is spontaneous, this is obtained in numerical work by using an initial condition with an axisymmetric solution and a three-dimensional perturbation superimposed on it.

High aspect ratio of two-disk flow family

As the aspect ratio $L > 1$ (1 < L < 4), the rotating disk acts as a pump, drawing in fluid axially and driving it away in an outward spiral. In a closed container, this fluid swirls along the cylindrical wall, spirals in across the stationary disk and then again turns into the axial direction towards the rotating disk. The inward spiralling motion results in an initial increase in swirl velocity, due to the conservation of angular momentum, and the creation of a concentrated vortex. For certain combinations of L and Re, this vortex undergoes breakdown as mentioned by Vogel (1968) [132]. A stagnation point followed by a recirculation zone of limited extent appears on the cylinder axis. Depending on the different aspect ratio L, different numbers of the breakdown arise at the swirl flow axis as shown in Figure 2.10, and again a transition diagram was obtained by Escudier (1984) [41] as shown in Figure 2.11. The vortex breakdown becomes unstable with increasing Re and L. Two unsteady regimes of the vortex breakdown were found. Above $L = 1.7$, for a certain Reynolds number, the stagnation point begins to oscillate periodically and axially, the threshold being represented by the symbol ∇. These oscillations were also recently studied numerically by Gelfgat *et al.* (2001) [49], Lopez *et al.* (2001) [77] and Stevens *et al.* (1999) [118]. It was found that the oscillation depends strongly

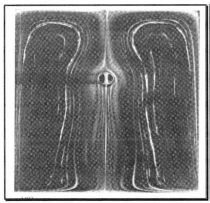

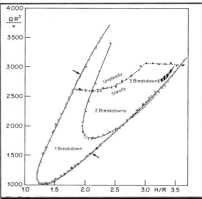

Figure 2.10: One example of vortex breakdown (one breakdown)

Figure 2.11: Transition diagram of vortex breakdown

on the azimuthal wave number k (for an axisymmetric circular roll $k = 0$). Three distinct oscillatory states exist in the region $L = 2.5$, among which the third is modulated by a very low-frequency oscillation (Stevens *et al.* [118]), and the connection between them is still unclear. However, the very low-frequency oscillation might be related to the second unsteady state of the vortex breakdown. For $L > 3.1$, as shown in the transition diagram, the first sign of a second unsteady motion of the vortex breakdown is a precession of the lower breakdown structure, symbolized by ▲.

In the most recent numerical study by Serre *et al.* (2001) [115], the spiral form transition of the vortex breakdown was connected with the precession of it in a DNS study for $L = 4$. At $Re = 2500$, the flow is axisymmetric and stationary without vortex breakdown. At $Re = 3700$, the flow undergoes a prolonged transient characterized by irregular oscillations before attaining a purely periodic behaviour of low frequency, $f/N = 0.11$, indicating the formation of vortex breakdown. If a non-axisymmetric perturbation is added at a given instant on a velocity component, the axisymmetry of the flow can be kept at a lower Reynolds number, $Re = 2500$. If the Reynolds number is increased to 6500, after a long enough time three-dimensional instabilities become visible

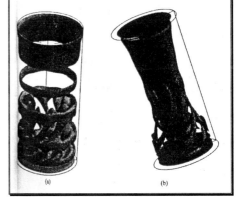

Figure 2.12: Iso-surface of the radial and azimuthal component of the vorticity

in the form of rolls convected from eight spiral arms (corresponding to the dominance of mode $k = 8$) by the three-dimensional main flow at about $z = H/3$ as shown in Figure 2.12. These structures travel downstream to the stationary disk layer and reach the flow region located between the side wall layer and the vortex breakdown region, where they combine into helical vortices. It was assumed that this phenomenon is associated with centrifugal instability. With the development of the instabilities, the breakdown mu-

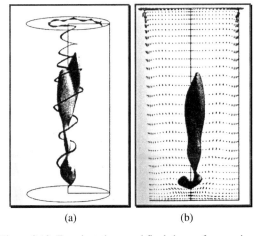

(a)   (b)

Figure 2.13: Transient shape and final shape of precessing swirls

tates from a bubble form in an elongated S-shape and gyrates about the centre as shown in Figure 2.13-a, also evidenced by the frequency decreasing from $f/N = 0.06$ to 0.033. Ultimately, it forms the final shape of the breakdown in precession as shown in Figure 2.13-b.

It can be concluded from the above introduction that axisymmetric and non-axisymmetric instabilities are inherent features of the rotational flow. These instabilities arise initially in the form of circular and spiral rolls, which evolve into helical or swirl flow in the centre of the rotation with increase in the aspect ratio, resulting in the formation of vortex breakdown. The vortex breakdown transits from steady into oscillatory unsteady states with certain levels of frequency and has a tendency to leap to a precessing motion for certain configurations, the frequency of which is characterized by a much smaller magnitude. In experimental work, no additional disturbances apart from geometry imperfections were needed in order to give rise to the instabilities, whereas in numerical work at least a point perturbation was introduced into the calculations.

2.1.8   PVC in swirl flows

Precessing motion takes place accompanying the occurrence of vortex breakdown. In flows where highly swirling flow ($S > 0.6$) is pronounced, e.g. cyclone chamber flows, swirl burner flows [55] or a reciprocating engine [88], the central vortex region of flow becomes unstable and starts to precess about the axis of symmetry. This phenomenon is termed "precessing vortex core" (PVC) (Gupta *et al.* [55]). A schematic representation of this motion is depicted in Figure 2.14. The PVC has a substantial influence on the swirl flow structure and has both positive and negative effects on the corresponding process, e.g. intensification of the turbu-

lence level in swirl generators, desta-
bilization of the flame, resonating and
vibration due to the low oscillations,
etc.

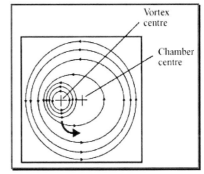

The PVC is manifested by oscillatory
motion superimposed on the basic
flow pattern. A typical velocity associated with the PVC is represented in
Figure 2.15. The velocity fluctuation
shows a fairly organized sinusoidal
wave shape instead of a steady line. A
series of non-dimensionalized parameters have been introduced by

Figure 2.14: Schematic representation of PVC in a
swirl flow chamber

many researchers to characterize the oscillation of the PVC in swirling flow:

- frequency parameter (Strouhal number):

$$Sh = f \cdot l / u, \qquad (2.23)$$

where f denotes the precession frequency and l and u the characteristic length and
velocity, respectively;

- swirl number (defined as in Section 2.1.5);

- Reynolds number.

It has been confirmed by different authors e.g. Chauaud (1965) [19], Gupta *et al.* [55] and
Alekseenko *et al.* (1999) [2] that the frequency increases linearly with swirl number in a moderate swirl number range and remains self-similar relative to the Reynolds number for a given
swirl number. Note that the above discussions are valid for well defined simple swirl flows. It
was pointed out by Nadarajah *et al.* (1998) [88] that many dominant frequencies arise associated with PVC in flow with complex geometries, e.g. in engines, preventing further characterization.

The spatial structure of the PVC
and the initial cause of it have
not been identified in spite of
numerous studies. Yazdabadi *et
al.* (1994) [143] obtained distributions for the velocity components over the entire cross-
section, employing LDA and the
phase averaging techniques.

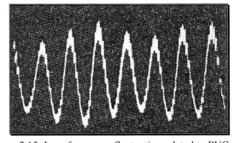

Figure 2.15: Low-frequency fluctuation related to PVC

They were able to locate the PVC offset quantitatively and to confirm the association of PVC with the counter flow region along the axis. The size of the offset increases with Reynolds number, but shrinks in size.

The above discussions show that the PVC is responsible for the unsteadiness in highly swirling flows and is characterized by periodic motion superimposed on the steady flow pattern or other unsteady flows such as turbulence. For well confined swirl flows, the frequency remains self-similar relative to the Reynolds number.

2.2 Turbulence and Stirring

Turbulence belongs to viscous instability. The critical Reynolds number characterizes the onset of instability due to inertial forces in the presence of viscous forces. Turbulence plays a positive role in the mixing process, therefore, in mixing processes a higher turbulence level is strived for in most cases. Turbulence can be induced in very different forms. In stirring processes, turbulence is introduced into the stirred vessels exclusively by the rotation of the stirrer element.

2.2.1 Turbulent flows and turbulence modelling

Turbulent motion is complex in nature, characterized by irregular, chaotic fluctuations. The velocity, pressure and temperature terms in the governing equations are random functions of time and space. From the engineering point of view, the statistical theory was introduced by dividing all instantaneous variables into an averaged value and a superimposed fluctuation. For velocities, an instantaneous velocity $U_i(t)$ can be decomposed as follows:

$$U_i(t) = \overline{U}_i + u_i(t) , \tag{2.24}$$

where $u_i(t)$ denotes the instantaneous deviation (fluctuation) from the time-averaged mean velocity $\overline{U}_i$, defined as

$$\overline{U}_i = \lim_{T \to \infty} \frac{1}{T} \int_{t_0}^{t_0+T} U_i(t)\mathrm{d}t . \tag{2.25}$$

The time-averaged fluctuations are zero according to

$$\overline{u_i} = \lim_{T \to \infty} \frac{1}{T} \int_{t_0}^{t_0+T} (U_i(t) - \overline{U}_i)\mathrm{d}t \equiv 0 . \tag{2.26}$$

In experimental and numerical investigations, the intensity of turbulence is defined as RMS values of the fluctuating velocities:

$$u_i' = \sqrt{\frac{1}{T} \int_{t_0}^{t_0+T} (U_i(t) - \bar{U}_i)^2 \, dt} \ . \tag{2.27}$$

Note that during the averaging process, the time scale T should be much larger than that of the fluctuation.

The governing equations of the time-averaged flow can be treated in the same way, which implies a separate averaging of each term. By replacing the instantaneous quantity with the mean and the fluctuating value in Equations (2.5) and (2.6) and keeping the density constant, the Reynolds averaged equations can be obtained for fluids with $\rho = $ constant and $\mu = $ constant [36]:

$$\frac{\partial \bar{U}_i}{\partial x_i} = 0 \ . \tag{2.28}$$

$$\rho \bar{U}_i \frac{\partial \bar{U}_j}{\partial x_i} = -\frac{\partial \bar{P}}{\partial x_j} + \frac{\partial}{\partial x_i} \left(\mu \frac{\partial \bar{U}_j}{\partial x_i} - \rho \overline{u_i u_j} \right) + F_j \ . \tag{2.29}$$

In the momentum equations, the underlined terms $\overline{u_i u_j}$ are known as the Reynolds stresses, and are actually inertia or momentum exchange effects. In statistical terminology the terms are referred to as correlations. Generally, there are six independent Reynolds unknowns. Owing to the existence of these new additional quantities, the equation system is no longer closed, since there are more unknowns than available equations. This is the well known "closure problem" of turbulent flows. In order to close the equation system, the Reynolds stress tensors have to be closed by turbulence modelling (RANS). Depending on the additional differential equations for turbulent quantities, such as Reynolds stresses, turbulent kinetic energy or length scale, the models are often classified into zero, one, two or other equations. More details are available elsewhere [139].

The most prevalent turbulence model in stirred flows is the so-called k-ε model developed in 1982 [57], a two-equation model with a transport equation for the turbulent kinetic energy k and its dissipation rate ε. More recent developments of k-ε models in stirred flows have been reported [134]. It was shown by Wechsler *et al.* [135, 136] that good agreement was achieved with regard to both velocities and trailing vortex structure in the vicinity of stirrer blades between numerical and experimental work, whereas considerable divations exist in the comparison of kinetic energy. This was attributed to the limitations of k-ε models in the prediction of swirling flows, which arise in the trailing vortex and in the recirculation zones [6]. Note that in most numerical investigations, the free surface was replaced by an imaginary lid for significant simplification in processing of grid and surface modelling. Only in the recent work of Ciafalo *et al.* (1996) [24] on unbaffled tanks and Serra *et al.* (2001) [114] on full-baffled tanks the free surface was simulated by moving grid and two-fluid model techniques, respectively.

An alternative to the time-averaged approach is the so-called direct numerical simulation (DNS), in which the Navier-Stokes equations are solved directly without any turbulence model based on resolving all turbulent scales up to the Kolmogorov scales. However, owing to limited computer resources, the DNS is preformed for very simple flows and serves as a basic tool for the development and verification of new turbulence models [11]. Bartels *et al.* (2000) [6] performed DNS to verify the predictions of the k-ε models in a stirred vessel geometrically identical with that of Wechsler [134], showing better agreement with the experimental results [108].

Large-eddy simulation (LES) bridges the gap between RANS and DNS by applying DNS to the large scales and RANS to the small scales [96]. Since the large eddies containing most of the energy are calculated directly, whereas the less important and more universal small eddies are easier to model, LES ensures simulation accuracy and efficiency. With the rapid developments in computer speed and capacities, LES is gaining entry into stirred flows [4, 30, 41].

2.2.2   Flow pattern in stirred vessels

Energy is introduced by the rotation of the stirrer element. The flow in stirred vessels is, first of all, a rotational flow. In the absence of baffles in stirred vessels, the fluid is moving primarily in a tangential direction. The fluid is put into solid rotation in the range of the stirrer blade and then possesses a free vortex structure up the tank wall. No great shear stress exists in the velocity profile, so little mixing takes place. A centre vortex is formed due to the pressure distribution as described in Section 2.1.6. As the depth of the centre vortex reaches the stirrer region, gas can be entrained into the whole vessel, leading to surface aeration.

In the presence of baffles, the secondary flow pattern is intensified owing to the redirection and disturbance effects of baffles. Depending on the impeller types, vessel geometries and installation configuration, two predominant secondary flow patterns exist in stirred flows embedded in the primary rotational flow: radial and axial flow. Correspondingly, the impeller types are classified into radial and axial pumping impellers, according to the pumping direction. Typical examples both in practice and in laboratories are the Rushton turbine (RT)

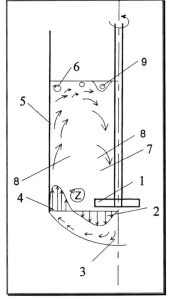

Figure 2.16: Circulation flow pattern of an axial pumping impeller [73]

for radial stirrers and the Pitched blade turbine (PBT) for axial stirrers. The secondary flow pattern is also often called the circulation pattern in stirred vessels and is considered to be of the great importance for the mixing process in stirred flows [152].

A typical circulation flow pattern of an axial pumping impeller is presented in Figure 2.16. In the impeller region (1), a discharge flow with a similar jet velocity distribution (2) is generated by the blade pumping effect. This jet meets its downstream stagnation point at the bottom (3) or sidewall of the cylinder. The jet flow is redirected by the collisions at the bottom. Thereafter, along the sidewall of the tank, a wall jet flow (4) is formed towards to the upper part of the tank. The wall jet flow decreases its intensity with increasing distance (5) from the stagnation point. The circulation pattern can seldom achieve the region near the surface (6). The jet flow is again redirected in this region and breaks up gradually. In region (7) the fluid is sucked back into the impeller owing to the conversation of mass. Near the central point of the circulation pattern (8), high turbulence exists despite the low velocities in this region. Note that, in spite of the presence of baffles, surface vortices (so-called "macro vortices" according to Liepe *et al.* [73]) appear in the surface region near the impeller shaft (9). It is acknowledged that such vortices are associated with the presence of baffles [73, 125, 129, 152].

2.2.3 Characteristics of stirred flows

Mechanical energy is introduced into the stirred vessel by the rotation of the stirrer elements and transported in the form of turbulent flow. Ultimately, the mechanical energy introduced will be transformed into heat under the effect of the viscosity of the stirred fluids. The flow status can be characterized by the Reynolds number as a ratio of forces as introduced in Section 2.1.2. For stirred flows, the stirrer Reynolds number is widely applied, defined as

$$Re = \frac{D \cdot (N \cdot D)}{\nu} = \frac{ND^2}{\nu}, \tag{2.30}$$

where the characteristic length is defined as the stirrer diameter D and the characteristic velocity as ND.

In a similar way, the Froude number characterizes the gravitational effects in stirred vessels with a free liquid surface and can be defined in stirred vessels as

$$Fr = \frac{(N \cdot D)^2}{Dg} = \frac{N^2 D}{g}. \tag{2.31}$$

The shape of the surface and, therefore, the flow pattern in the vessel are affected by the gravitational field. This is particularly noticeable in unbaffled tanks where a large centre vortex occur. The vortex depth can be related to the Froude number as described by Bird *et al.* (1960) [7].

The energy consumption of a stirrer system can be characterized by the power number Po:

$$Po = \frac{P}{\rho N^3 D^5} ,$$
(2.32)

where P denotes the power needed. As demonstrated by Uhl *et al.* (1966) [129], the power number is a pressure coefficient and represents the ratio of pressure differences producing flow to inertial forces. Power P is fed to overcome the drag forces over the surface of the impeller blades. The power number is usually given as a function of the impeller Reynolds number and the impeller and tank geometry. For fully baffled tanks, the flow is considered fully turbulent in the impeller region of the stirred tank when the power number becomes constant despite increasing Reynolds number (Rushton, *et al.* 1954 [106]).

A scale reflecting the pumping capacity of the stirrer and the circulation intensity is the flow number. The primary flow number Fl is widely applied in stirred flows:

$$Fl = \frac{Q}{ND^3} ,$$
(2.33)

where Q represents the volume flow rate discharged directly from the stirrer blades. The pumping effect of the stirrer blades can be viewed as an imaginary pump. As the discharged flow departs from the impeller region, additionally the surrounding fluids will be brought into the discharge flow jet by "entrainment". Therefore, the pumped fluid volume increases with increasing distance from the impeller. Normally, taking this entrainment effect into account, the so-called secondary flow number is introduced by replacing Q with Q_c, which considered to be able to characterize the impeller pumping efficiency more quantitatively. However, it was shown by Yianneskis (2000) [144] and Schäfer (2001) [108] that the secondary flow number is difficult to determine owing to the very varied locations of the velocity profiles.

Normally, up to four baffles are installed in stirred vessels in order to weaken the ineffective tangential flow and intensify the large-scale turbulent circulations. The influence of baffles can be characterized by the baffle intensity number Bi ("Bewehrungszahl" according to Liepe *et al.*, 1998 [73]). For a one-stirrer system with a flat-bottomed tank, the baffle intensity number is

$$Bi = n_b^{0.8} \frac{W_b}{T} \frac{H_b}{T} .$$
(2.34)

The baffle status can be classified into three levels according to the baffle intensity numbers:

- Unbaffled or inefficiently baffled: strong tangential velocity still exists, the power number remains low and the central vortex is fully developed. Note that for highly viscous fluids or flows with very low Reynolds numbers, the application of baffles is useless since the flow is pure laminar and dead zones can appear, leading to bad mixing quality. This is also called the self-baffled effect according to Liepe *et al.* [73].

- Partially baffled: the central vortex is considerably suppressed, and instead of the central vortex, surface vortices arise, resulting in air entrainment at substantial speed.

- Fully baffled: the power number reaches the maximum, the surface vortices are suppressed and the central vortex disappears totally.

Based on extensive experiments by Liepe *et al.* [73], the critical baffle intensity numbers were determined. For RT, this number depends on the power number:

$$Bi_c = 0.36 Po^{1/3} \frac{D}{T}.$$ (2.35)

2.2.4 Turbulence distribution in stirred vessels

Turbulent mixing occurs at two different levels: macromixing and micromixing. As the names imply, macromixing is large-scale or full-tank mixing comprising convection, turbulent diffusion and turbulence eddy break-up, whereas micromixing represents the mixing process corresponding to the smallest turbulence scale [108, 152]. Turbulence is generated by the rotation of the impeller with a macro length scale corresponding to the stirrer blade size, and the turbulence kinetic energy is transported by convection into eddies with smaller and smaller scales down to the Kolmogorov micro-scale.

The flow and turbulence parameters are strongly dependent on the locations in stirred vessels, while the impeller Reynolds number defined as in Section 2.2.3 characterizes only the discharge flow in the direct vicinity of impellers. In general, it is useful to distinguish processes which are affected by turbulence close to the stirrer, for example, power input, maximum local dissipation, maximum of the shear stress etc, and processes which are determined by turbulence far from the stirrer, e.g. surface turbulence and wall jet flow break-up. It is obvious that for surface aeration the turbulence far from the stirrer plays a more essential role.

Based on the impeller Reynolds number, the stirred flow can be divided into four levels:

- $Re > Re_{ft}$: *ft* denotes fully developed turbulence in the whole tank except at the wall boundary. The Reynolds number and hence the viscosity do not influence the flow and turbulence field. The velocity and turbulence quantities obey the scale-up rules well.

- $Re_{sb} < Re < Re_{ft}$: *sb* represents the self-baffled state. The flow is still turbulent but not fully developed. However, a self-baffled state is still not achieved.

- $Re_{st} < Re < Re_{sb}$: *st* means steeping flow. The application of baffles is unnecessary.

- $Re < Re_{st}$: pure steeping flow, no mixing happens.

The critical Reynolds number for the above-mentioned levels can be determined by experimental investigations according to different types of stirrers and geometry configurations and also different flow regions of interest. A rough estimation for RT and PBT in baffled tanks was given by Liepe *et al.* [73] and is listed in Table 2.2. It is obvious that much higher Reynolds numbers will be needed if a fully developed turbulence, even in the region far from the impeller region, is strived for.

Stirrer element	Re_{sb}	Re_{ft} close to stirrer	Re_{ft} far from stirrer
RT	250	2×10^4	5×10^5
PBT	700	2.5×10^4	4×10^5

Table 2.2: Critical Reynolds numbers

2.3   Air Entrainment

2.3.1   Introduction

Free-surface flows can be defined as a liquid phase flowing with a "pseudo-continuous" interface with the atmosphere or with an air-filled cavity. Air entrainment over the free surface is defined as the entrainment/entrapment of undissolved air bubbles and air pockets that are carried away with the flowing fluid (see Chanson, 1996 [22]). It often occurs in free-surface flows existing both in nature and in industrial applications. Typical examples are plunging jets, hydraulic jumps, open channels, etc. Figure 2.17 represents schematically examples of free-surface aeration in different flow situations. In the absence of air entrainment, the air-water interface is well marked and continuous. If air is entrained into the fluid, the exact location of the interface between the flowing fluid and the air phase becomes undetermined.

2.3.2   Air entrainment mechanisms

The mechanisms of air entrainment differ substantially between various flow features and individual flow conditions. In each case several explanations have been proposed to describe the mechanisms of free-surface aeration.

With plunging water jets (including hydraulic jump flows), air entrainment results from some discontinuity at the impingement point. Two types of mechanisms are described by Chanson and Cummings (1994) [21]: individual air bubble entrainment by low-velocity jets, and a mechanism of ventilated air sheets and re-entrant jets for high-velocity jets. Initial aeration of the impinging jet free surface may feed back the entrainment. Downstream of the entrainment point, air bubbles are carried away by turbulent vortices with the main axis perpendicular to the flow direction. In open-channel flows, the free-surface wave instability and turbulent ve-

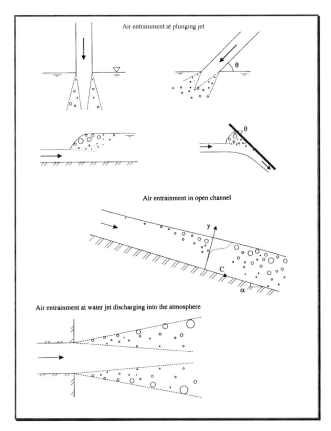

Figure 2.17: Examples of free-surface aeration

locity fluctuations are responsible for air entrainment according to Hinze (1961) [61]. In jet flows with discharging of dense fluid into a lighter medium (e.g. a water jet discharging into air), air is entrapped at free-surface irregularities and diffused within the jet by turbulent velocity fluctuations normal to the mean flow direction. However, it is well recognised that free-surface entrainment occurs when the turbulence level is large enough to overcome both surface tension and buoyancy effects. More details about the state of the art of air entrainment can be found elsewhere [22].

Analytical and numerical studies of air-water flow are particularly complex because of the large number of relevant equations. Experimental investigations are also difficult despite the rapid development of flow measuring techniques. Research on this subject is limited in both dimensional analysis and similitude studies. The most important parameters are given in the following dimensionless numbers: the geometrical parameters (ratios), the Froude number Fr, the Reynolds number Re, the Weber number We, the turbulent intensity Tu and the characteristic bubble size.

3   EXPERIMENTAL AND NUMERICAL TECHNIQUES EMPLOYED

In this chapter, experimental and numerical methods in fluid mechanics which were employed in the present work are introduced. Especially experimental techniques, including flow visualization, laser Doppler anemometry, were mainly applied for the experimental studies both in stirred vessels and in simplified geometry configurations. With the help of the fast developing numerical methods, flow investigations have been developed in a promising direction. A combination of numerical and experimental methods has inspired research work related to fluid mechanics. The basis of the numerical methods employed in the present work is introduced below.

3.1   Laser Light Sheet Visualization and Video Analysis Technique

The simplest method to observe the flow field is the tracer method, based on seeding of particles following the flow motion. The laser sheet visualization technique is one of the most common tracer methods. In comparison with LDA and PIV measuring techniques, the tracer methods have the shortcoming of poorly quantitative evaluation. However, these methods permit a quick overview of and insight into the flow field of interest simply and economically in terms of both equipment costs and measuring time. Especially for unsteady flow phenomena, tracer methods find wide application owing to their ability to obtain a macroscopic instantaneous flow pattern (Gad-el-Hek, 1988 [47]). Among the tracer methods, laser sheet visualization technique permits observations in a two-dimensional flow field. It was introduced into stirred flows by Rushton and Sachs (1954) [106]. Metzner and Taylor (1960) [85] obtained the flow pattern including the horizontal flow field in a tank with laser sheet visualization by using a transparent vessel bottom. In the present work, the instantaneous flow structure and flow phenomena associated with instabilities in rotating disk flow were investigated mainly with help of the laser light sheet visualization technique.

The laser light sheet visualization technique is based on the illumination of a light plane in which the flow domain of interest is involved. Seeded particles following the flow through this plane reflect the light and thus a two-dimensional picture of the fluid motion can be visualized. If needed, a three-dimensional flow field can likewise be acquired by successful illumination in more planes [63]. Depending on the size of the measuring section and type of flow, normal light or laser light can be used as the light source. For normal light, a split diaphragm is often used to generate a light plane. In practice, laser light is more often used, since a laser can supply light with higher intensity. The laser beam can be scattered in a plane with a cylindrical lens or a swinging mirror. The selection of particles reflecting light is of great importance, since it must follow the flow well and at the same time reflect sufficient light intensity.

In the present work, laser light sheet visualization was fully available for the rotating flow investigations. As the laser light source of the laser sheet equipment, a 170 mW mono-mode laser diode ("Bramlas") operating at a wavelength of 532 nm was used, allowing a compact experimental set-up thanks to its small size. The laser diode is located in an optical frame together with the optic lens system. The laser beam is collimated and focused by the optical system at the central point of the swinging mirror. The swinging motion of the mirror is controlled by a function generator (TT*i*, TG215) and driven by a current amplifier developed at LSTM-Erlangen. A sinusoidal pulse working in the range 20 - 30 Hz was selected to obtain a satisfactory picture quality. As tracer particles, aluminium powder was selected for flow investigations in the present work. A detailed set-up description of the laser sheet equipment can be found elsewhere [73].

For recording the visualization result, pictures were taken with a Canon T90 single-reflex camera with a Kodak Tmax 400 negative film (sensitivity ASA 3200). The exposure time was about 1 s. It should be noted that the flow phenomena associated with the instabilities were very often concealed in the photographs captured by the camera owing to the relatively long exposure time needed for an acceptable picture quality. Therefore, for instability analysis, the laser light sheet visualization was also recorded with a Panasonic CCD video camera. The videos were then edited and transferred to a PC with the help of an AV MASTER 2000 interface card (FAST Multimedia AG) which was installed in the PC, allowing detailed analysis on a PC frame by frame.

3.2   Laser Doppler Anemometry

Laser Doppler anemometry (LDA) has proved to be more accurate in the measurement of flow fields in stirred vessels than any other flow measuring technique, thanks to its main advantages of high accuracy, non-intrusive nature and fast response, which is of great importance for strongly unsteady and highly turbulent flow.

3.2.1   Basic principle of LDA

The most often used LDA is a "one-component dual-beam" LDA system, where "dual" means two beams, often realized by splitting one laser beam into two beams, and "one-component" indicates that simultaneously only one component of the velocity in space can be measured. The principle can be explained by an interference model as shown in Figure 3.1. Two coherent laser beams intersect at a point where they are focused with an angle 2φ and hence form interference with dark and light fringes which are perpendicular to the moving direction of a particle. This range is normally called the measuring control volume. The fringe spacing can be calculated as

$$\Delta x = \frac{\lambda}{2\sin\varphi},\tag{3.1}$$

where λ represents the wavelength of the laser beam.

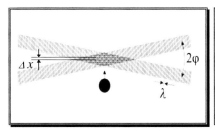
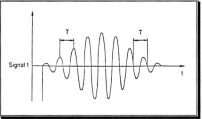

Figure 3.1: LDA interference model   Figure 3.2: Typical Doppler burst

The moving particle in the control volume reflects the interference corresponding to the light intensity distribution. The light scattered by the particle is modulated in intensity with a Doppler frequency f_D and one period $1/f_D$ corresponds to the time which a particle needs to pass through the control volume with a perpendicular velocity component $u_\perp$. This relation is described by

$$u_\perp = \frac{\lambda}{2\sin\varphi} f_D. \tag{3.2}$$

The low frequency (burst) is generated by the Gaussian distribution of the beam intensity and can be filtered out by a high-pass frequency filter. A typical Doppler burst after filtering is shown in Figure 3.2. From the linear relation between f_D and $u_\perp$, the velocity information can be obtained, and no calibration is needed. Detailed background of the principle and practices of LDA can be found in Durst (1976) [38] and Ruck (1987) [105].

3.2.2 LDA system employed

3.2.2.1 Experimental set-up

At LSTM-Erlangen, a multifunctional stirrer LDA test rig has been developed within the framework of several research projects. The set-up of the test rig is represented in Figure 3.3. The basic structure consists of three main subsystems: the measuring section, the traversing equipment and the LDA measurement system.

3.2.2.2 Measurement section

The LDA experiments were carried out in two flat-bottomed, cylindrical transparent Duran-glass vessels of diameter $T = 152$ and 400 mm. The liquid height in the vessel was set to $H = T$. The geometry of the mixing vessel and the coordinate system is schematically illustrated in Figure 3.4. The main geometry parameters of the vessels are listed in Table 3.1.

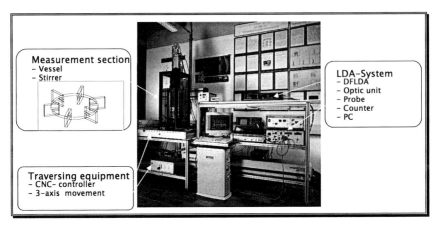

Figure 3.3: LDA experimental set-up

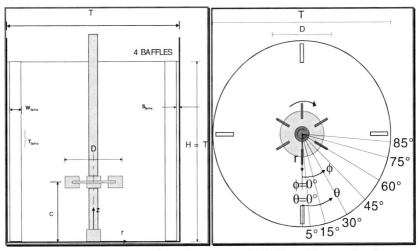

(a): Schematic representation of vessels   (b) Baffle positions employed

Figure 3.4: Arrangement of measurement section

	T	C	H	Baffles		
				Width	Wall distance	Thickness
Vessel I	152	50	152	15	2.6	2
Vessel II	400	133.3	400	30	10	5

Table 3.1: Geometric dimensions (mm) of the vessels employed

The baffle numbers could be varied between 2 and 4 during the investigations, corresponding to baffle intensity numbers of 0.174 and 0.303, respectively, according to Equation (2.34). For a standard RT with $Po = 4.48$ (see Schäfer, 2001 [108]) as employed in the present work, according to Equation 2.31, the critical baffle intensity number was 0.199. Therefore only the four wall baffles ensure a fully baffled configuration. In comparison, the critical baffle intensity number for PBT with $Po = 1.44$ (see also Schäfer [108]) was 0.174 , indicating that a two-baffle configuration achieves the fully baffled status. The baffles are located symmetrically in the circumference direction, each having a gap of 0.025 T to the vessel wall. The baffles are fixed to a holding device which can be rotated to any angles with a resolution of 5°, enabling measurements to be formed at different vertical planes between baffles as shown in Figure 3.4-b.

The cylindrical vessel was located in a square tank made of Duran glass, in order to eliminate the distorting effect of the rounded surface of the cylindrical vessel through which the laser beams have to pass. Both the cylindrical vessel itself and the gap between the cylindrical vessel and the square tank were filled with the working fluid, a silicone oil mixture of AN20 and AS4 (WACKER) with a mixing ratio of 65.53:34.47. This mixture has the same refractive index as the Duran glass of the vessels ($n = 1.468$) in the working wavelength range of the laser beams employed at an operating temperature of 21°C. Since the refractive indices of Duran glass and of the working fluid vary differently with the temperature, the temperature was kept constant at 21°C during the measurements by a water cooling system installed at the bottom of the cylindrical vessel and monitored by a thermometer inserted in the cylindrical mixing vessel. This refractive index-matched measuring configuration permits automatic measurements. The dynamic viscosity of the silicone oil mixture at 21°C is 16.4 mPas and the density is 1021 kg/m^3.

The geometry of the stirrer elements is depicted graphically in Figure 3.5-a for RT and in Figure 3.5-b for PBT. The dimensions of the stirrers will be described in detail in the corresponding sections. If not mentioned, the off-bottom clearance C was set to $T/3$ for both the

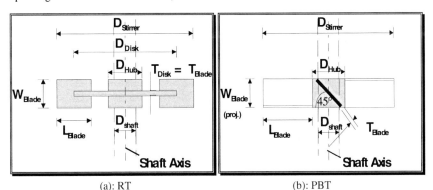

(a): RT   (b): PBT

Figure 3.5: Geometry of stirrers

rotating disk and the stirrers.

3.2.2.3   LDA measuring system and data acquisition

The LDA system employed in the present work was a diode fibre laser Doppler anemometer (DFLDA), which was developed at LSTM-Erlangen and worked in the backward scattering mode. In comparison with the forward scattering mode, the main disadvantage is the much lower signal rate due to the ca. 1000 times lower scattered light intensity [38]. However, together with the use of a laser diode and fibre optic probes, it offers considerable advantages, e.g. flexibility, compactness and easy handling of optics, allowing easier traversing and automatic measurement operation. In addition, the backward scattering mode is the only appropriate way to obtain the tangential velocity component, whereas in the forward scattering mode an offset from the vessel axis is needed to allow the laser beams to pass the impeller shaft. A significant difference in measured velocity appears if the offset exceeds a certain range [67]. The disadvantage due to the low scatter intensity in the backward scattering mode could be compensated for the present work by seeding into the working fluid silver-coated hollow glass spheres (DANTEC-INVENT). The average sphere diameter is 10 µm, ensuring satisfactory flow following characteristics. A measuring rate in the range 60-400 Hz was obtained depending on the measured positions in the vessels and a sufficient signal-to-noise ratios.

The compact basic unit of the DFLDA system, as shown in Figure 3.6, comprised all optical components, e.g. laser diode, the fibre in-coupling unit and the avalanche photodiode (APD) for signal detection. The laser source is a 100 mW mono-mode laser diode (LD) operating at 832 nm and packaged with temperature regulation and correction optics for the astigmatism effect and for achieving a circular beam profile. The output power of this package is 94 mW and the beam is nearly circular with a diameter of 5.6 mm. This beam is collimated to a di-

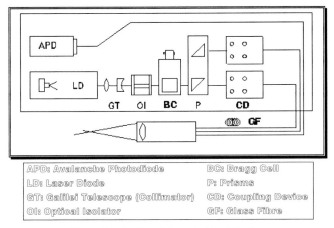

Figure 3.6: Optic unit layout of the DFLDA

ameter of 1.2 mm with the beam waist positioned at the front face of the SELFOC lenses that were used for in-coupling (CD). A passive optical isolator (OI) is used to suppress reflections back into the laser diode and thus external resonator effects. The beam is split into two beams of equal intensity using the zero- and first-order beams exiting from a Bragg cell (BC). Whereas the first beam is directed into the in-coupling device (CD) directly, two prisms deflect the second beam on to the second in-coupling device. Transformed by glass fibres (GF) ending in the probe head, the two beams are focused and form a measuring control volume by an exchangeable lens permitting high flexibility in positioning the control volume. The light scattered by seeding particles moving through the measuring volume is collected by a receiving optics in the backward direction and then directed back into the basic unit by multi-mode glass fibre, and ultimately focused on an avalanche photodiode (APD). The shift frequency resulting from the Bragg cell (BC) operating at about 60 MHz is decreased by a down-mixer to the required shift frequency ranges, in order to match different flow velocity ranges. More details about the DFLDA system were described in by Stiegmeier and Tropea (1992) [119]. The specific parameters of the DFLDA system used for the present study are listed in Table 3.2.

The LDA signals collected by the APD are processed by a frequency counter (TSI Model 1980B). The signal from the APD is filtered to remove low-frequency pedestal and high-frequency noise, and consequently converted into the Doppler frequency and velocities. Finally, the measured messages are read into a PC through an LDA interface card (DOSTEK) for restoring and later analysis. The interface is controlled by a computer program to acquire

Item	LDA
Wave length	832 nm
Focus of front lens	200 mm
Focus of collection lens	-
Focus of receiving lens	60 mm
Beam distance	16 mm
Half angle of beams (in air)	2.29°
Beam diameter	3.5 mm
Fringe number (without shift)	6
Fringe space in air	10.49 µm
Measuring volume diameter (air)	61 µm
Measuring volume length (air)	1.53 mm

Table 3.2: LDA configuration parameters

data from the counter depending on either the arrival time of the LDA signals or the angle information of the angle encoder mounted on the top of the motor, namely time series of velocities or angle-resolved measurement, respectively. These two types of measurements can be switched between each other by the adjustment of jumpers on the interface card. The angle-resolved measurement is especially appropriate for the flow in the vicinity of the impeller blades and details on this type of measurement are described in [108]. In the present work, the time series of the velocities were more often measured for macro-instability frequency transformation analysis. The time-averaged velocities $\bar{U}$ and the RMS values u' are defined by the following equations:

$$\bar{U} = \frac{\sum_{i=2}^{N} \frac{1}{2}\left(U_i(t_i) + U_{i-1}(t_{i-1})\right) \cdot \Delta t_i}{\sum_{i=2}^{N} \Delta t_i}, \tag{3.3}$$

$$u' = \sqrt{\frac{\sum_{i=2}^{N}\left(U_i(t_i) - \bar{U}\right)^2 \cdot \Delta t_i}{\sum_{i=2}^{N} \Delta t_i}}. \tag{3.4}$$

In the DFLDA measuring system employed, automatic measurements were performed by using a 3D traversing system on which the LDA probe is mounted. The traversing system is driven by three step motors and controlled by a CNC system connected with a PC. The accuracy in traversing in three directions is 0.1 mm.

In order to have an overview of the turbulent cylindrical stirred flows, all three components of the velocities and the RMS values, namely the radial, tangential and axial components, were measured at the same spatial point but not at the same time by the one-component LDA system. The measurements were based on the assumption that the flow at constant rotational speed is periodically steady. In addition, the axisymmetry of the stirred vessel geometry simplified the performance of three-component measurements. Figure 3.7 illustrates schematically the three components measurement. As shown in Figure 3.7-(a), under the assumption of periodic steadiness of the stirred flow, the radial and the tangential components were measured at the same point but in two different vertical planes. By rotating the measuring probe by 90° as illustrated in Figure 3.7-(b), the axial component measurement can be carried out.

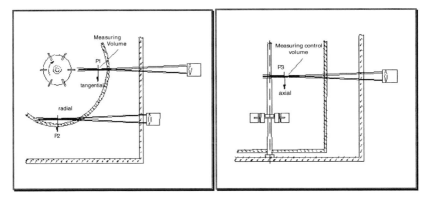

(a): Tangential and radial components   (b): Axial component

Figure 3.7: Three-component measurement of the velocity

3.2.2.4 Sample size and error analysis

There is a series of error sources in the velocity measurements with LDA, and the accuracy of the measurements varies with the measuring positions. A detailed discussion of these error sources from the measuring instruments was given in [38]. The statistical error in averaging the measured values also influences the accuracy. To eliminate or reduce this type of error, a sufficient sampling size for one measurement, namely the discrete sampling numbers of the velocity, is needed. Assuming that single measuring values have a Gaussian distribution at one fixed measuring point, the statistical error in the averaging processing can be estimated in percentage by

$$\varepsilon_{\bar{U}} = z \cdot \frac{\left(u'/\bar{U}\right)}{\sqrt{N}},\tag{3.5}$$

where z represents a given confidence interval and normally it was set $z = 1.96$, corresponding to an accuracy criterion of 95%, which means that 95% of the values are in the confidence interval range. N is the statistically independent sample number. The statistical error can be expressed with the RMS value defined in Equation (3.4). In turbulent flow measurements, the ratio of the RMS value to the averaged values as expressed in Equation (3.3) is normally defined as the turbulence intensity:

$$Tu = \frac{u'}{\bar{U}}.\tag{3.6}$$

The error of the RMS value can be estimated as

$$\varepsilon_{rms} = \frac{z}{\sqrt{2N}}.\tag{3.7}$$

The turbulence intensity in the stirred vessel can be estimated from the probe measurements and a maximum of 40% was obtained in the stirrer region. The necessary sampling numbers can then be estimated:

$$N = \frac{z^2 \cdot Tu^2}{\varepsilon_{\bar{U}}^2} = \frac{(1.96)^2 \cdot (0.4)^2}{(0.03)^2} = 683.$$ (3.8)

If an impeller with diameter D rotates with a rotational speed of 300 rpm, the integral time scale is estimated with characteristic length and velocity scale l and v as

$$I_t = \frac{l}{v} = \frac{D}{\pi DN} = \frac{1}{\pi N} = 6.37 \times 10^{-2}.$$ (3.9)

which corresponds to a statistically independent sampling rate of 7.85 Hz, while during the measurement a 10 ~ 100-fold higher sampling rate was achieved for an optimised signal quality. The sampling number should be of the order of at least 10 times larger than 683. For more safety and for easier measuring and evaluation programs, the sampling size was selected as $N = 10,000$. For macro instability dominant frequency analysis, the sampling time has the most important influence on the frequency transformation in lower frequency range. This will be discussed in Section 6.2.2. The sampling number was increased for corresponding measurements.

3.3   Numeric Method and Flow Computation

3.3.1   General introduction

The incompressible flows involved are governed by a system of non-linear Navier-Stokes equations as presented in Chapter 2. Analytical solutions can be obtained exclusively for a few limited cases. Therefore, numerical simulation is undergoing significant expansion in practice. The most commonly used numerical methods for computations in fluid dynamics are the finite difference (FD), finite volume (FV) and finite elements (FEM) methods.

Generally, each numerical method needs to use a discretized domain named grid to simulate the physical domain and the approximate solution can be obtained from the grid points. For flows in irregular complex geometries which appear more often in practical applications, the choice of coordinate systems is of great importance. Curvilinear non-orthogonal coordinates can follow the geometry of the solid boundaries limiting the flow field, and the grid lines may intersect at arbitrary angles, and therefore it is appropriate to simulate the flow in complex geometries, e.g., the rotating disk flow in a rectangular tank considered in the present work. In this case, the equations can be transformed from the physical domain to curvilinear, non-orthogonal coordinates by a reasonable numerical discretization, the so-called computational domain.

In the present work, the finite volume method was used to accomplish the computation, while an implicit method based on the SIMPLE algorithm [93] was applied for velocity-pressure coupling to solve the generated linearized equations. For more information, see Ferziger and Perić (1996) [45].

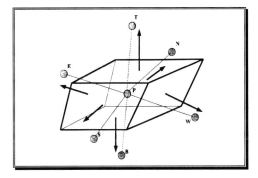

Figure 3.8: A general 3D control volume

The computation employed in the present work is based on a CFD code, FASTEST-3D [35], which was developed at LSTM-Erlangen. The grid generation procedure was performed in IGG, commercial grid generator software from NUMECA [129].

3.3.2   Finite volume method

3.3.2.1 Discretization in space

In order to discretize the governing equations, the physical domain is divided into finite control volumes (CVs). Each control volume is surrounding one central node P as shown in Figure 3.8. The approximate value of the solution is represented by the central node inside the control volume. Based on the Gauss divergence theorem, the conservation equations are integrated over the control volume, leading to an equation containing convective and diffusive fluxes through the control volume faces, volume integrated sources and a volume discretized transient term. The general transport equation for a quantity is

$$\frac{\partial}{\partial t}\left(\rho\Phi\right)+\frac{\partial}{\partial x_j}\left(\rho U_j\Phi-\Gamma_\Phi\frac{\partial\Phi}{\partial x_j}\right)=s_\Phi,\tag{3.10}$$

where Φ represents the transported quantity, Γ_Φ the diffusion coefficient and s_Φ the general source term. Integrated over a control volume, the above equation can be expressed as

$$\frac{\partial}{\partial t}\int_V\rho\Phi dV+\int_A\left(\rho U_i\Phi-\Gamma_\Phi\frac{\partial\Phi}{\partial x_i}\right)dA_i=\int_V s_\Phi dV,\tag{3.11}$$

where V represents the volume of CV and A_i is the surface vector enclosing the volume V. The two surface terms represent the convective and diffusive fluxes, respectively. For a hexahedral control volume surrounding the central node P shown in Figure 3.8, the convective and diffusive fluxes are summarized over volume faces east, west, north, south, top and bottom, which are connected with the neighbours nb (symbolized as E, W, N, S, T and B, respectively):

$$\int_A \left(\rho U_i \Phi - \Gamma_\Phi \frac{\partial \Phi}{\partial x_i} \right) dA_i = \sum_{nb} \left(\rho U_i \Phi - \Gamma_\Phi \frac{\partial \Phi}{\partial x_i} \right) A_i. \tag{3.12}$$

According to the continuity equation, the sum of fluxes F_{nb} going through all faces of the control volume satisfies the equation

$$F_e - F_w + F_n - F_s + F_t - F_b = S_\Phi. \tag{3.13}$$

The approximate values at the surfaces of each CV are calculated by linear interpolation with the node values. The diffusion fluxes can be approximated by the midpoint rule, and the gradients can be approximated by the central differencing scheme (CDS). For convection fluxes, a combined approximation from two discretization schemes is often applied in the following way:

$$F = F^{UDS} + \gamma(F^{CDS} - F^{UDS}), \tag{3.14}$$

where UDS denotes the upwind differencing scheme and γ the blending factor. With CDS a second-order accuracy can be obtained, whereas UDS provides only first-order accuracy in space, but is unconditionally stable.

By summing all the flux approximations and source terms for one CV in the discretization process, an algebraic equation is obtained which relates the variable value at the centre of the CV to the values at several neighbouring CVs in the following way:

$$A_P \Phi_P = \sum_{nb} A_{nb} \Phi_{nb} + s_\Phi, \quad nb = E, W, N, S, T, B, \tag{3.15}$$

where A_P and A_{nb} represent the coefficients at the node of the CV and of the neighbouring CVs. For the complete problem with m CVs, a coefficient matrix A_m is obtained:

$$A_m(\Phi_m) = s_m. \tag{3.16}$$

3.3.2.2 Discretization in time

For unsteady problems, a fourth coordinate direction, namely time, has to be introduced in the discretization. Among different discretization schemes, resulting in different accuracy, three schemes are commonly used: the *explicit Euler* method, *implicit Euler* method and *Crank-Nicolson* method. After a spatial integration of the transient term in Equation (3.10), the discretization results:

$$\frac{\left(\rho\Phi_p V\right)^{n+1}-\left(\rho\Phi_p V\right)^{n}}{\Delta t}=\theta\left(A_P\Phi_p+\sum_{nb}A_{nb}\Phi_{nb}+s_P\right)^{n+1}$$
$$+(1-\theta)\left(A_P\Phi_p+\sum_{nb}A_{nb}\Phi_{nb}+s_P\right)^{n}. \tag{3.17}$$

With θ equals 0, 1/2 and 1, the equation represents the first-order explicit Euler, the second-order Crank-Nicolson and the first-order implicit Euler time discretization, respectively. The Crank-Nicolson method requires more memory than the other two, since all variables have to be saved at two time levels. It is unconditionally stable and second-order accurate. The Euler explicit and implicit methods correspond to a forward and backward differencing scheme, respectively. The stability of the Euler explicit method is limited by the Courant number:

$$Co=\frac{u\Delta t}{\Delta x}\leq 1, \tag{3.18}$$

where Δx is the cell spacing.

3.3.3 Grid generation

A grid with discrete points serves as the discretization of the flow domain of interest. The grid classification mainly distinguishes between structured and unstructured grids. Both approaches are suitable for complex geometry. The generation of an unstructured grid can be easily automated, whereas the structured grid generation can be a very challenging task in practice. Since the accuracy, the rate of convergence and stability may depend very sensitively on the numerical grids, the choice of the grid type and the grid point distribution is of great importance.

The numerical analysis of practical flows normally has complex geometries. In order to retain the advantage of structured grids, complex regions can be decomposed into separate blocks and each sub-block can be resolved with a

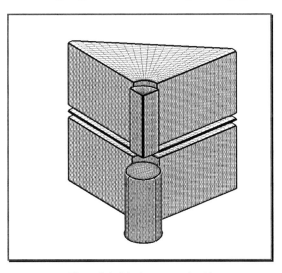

Figure 3.9: Block structured grid

structured grid. Figure 3.9 represents schematically an example (rotating disk in a square tank) of a decomposed flow region.

When block-structured grids are used, all variables have to be exchanged between the blocks at the boundaries, which provides this information as boundary conditions in the neighbouring blocks. In addition, most practical cases are somehow symmetric with respect to geometry. The grid generation is closely involved with these symmetries, since the flow domain of interest can be calculated in a part of the whole geometry, resulting in a reduction in the number of CVs and thus the calculation time.

4 INTRODUCTORY STUDIES OF ROTATING DISK FLOWS

Stirred flows are, first of all, rotational flows, since the initial motion is generated by the rotation of the stirrer elements. Owing to the effect of centrifugal forces, viscous forces, Coriolis forces, gravity and so on, rotational flows normally form a secondary circulation pattern which can be characterized by turbulence, micro-scale dissipation and macro-scale instabilities [79]. The common features of rotational flows and the instabilities in the rotating disk flow were summarized in Section 2.1. Generally, different types of the spiral and circular instabilities can take place in the boundary layer of the disk. With higher L/D ratios, vortex breakdown appears in the centre of the rotating disk flow, often accompanied by precessing motions of the vortex core (PVC). However, most of the instability investigations on rotating disk flows were limited to a confined set-up which was closed with two disks and a cylindrical sidewall. This arrangement is different from that of normal stirred flows. Contrary to the literature presented in Section 2.1, the flow induced by a rotating disk in a square tank was investigated analytically, numerically and experimentally in the present work, which combines the properties discussed in the literature with stirred flows. Finally, in order to associate the rotating disk flow and stirred flows, the rotating disk flow in a baffled tank was measured and compared with the flow induced by a standard radial impeller RT in the same tank.

4.1 Geometry Set-up

Square tanks can suppress the solid body motion, and hence are applied as moderate baffled tanks in stirring techniques [125]. The square tank applied in the present work was made of glass for visual observations. It had a length of 500 mm, and was filled with water to a filling height of 400 mm. The diameter of the disk was 100 mm and the thickness 1.5 mm, which was relatively small in comparison with the tank. The disk was mounted in the centre of the square tank and had an off-bottom clearance of 200 mm. In order to minimize the influences of the shaft of the rotating disk on the instabilities, the shaft diameter was chosen as 5 mm. A DC-motor was mounted above the square tank to drive the disk. The geometry is illustrated in Figure 4.1.

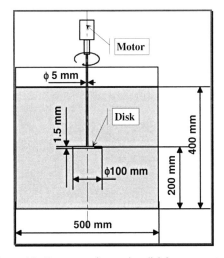

Figure 4.1: Geometry of a rotating disk in a square tank

4.2   Analytical Investigation of Rotating Disk Flow

4.2.1   Introduction

A very attractive feature of rotating disk flow is the exact solutions of the Navier-Stokes equations in the disk boundary layer. Exact solutions of the Navier-Stokes equations are generally extremely difficult owing to the nonlinearity of these equations. Despite this, there are some special cases where exact solutions can be obtained for incompressible flow. The so-called von Kármán boundary layer flow induced by a rotating disk is an example and was described in detail in [112]. The disk is modelled as an infinite planar disk rotating at a constant angular velocity ω about an axis perpendicular to its plane in a fluid, which is otherwise at rest. Because of the no-slip condition and the viscosity, the layer of the fluid directly at the disk is carried along and driven outwards by the centrifugal force. New fluid particles are then continuously pulled on to the disk in the axial direction and then ejected centrifugally again. This results in a secondary flow beside the primary rotating motion as mentioned in Section 2.1.4. Figure 4.2 shows this flow in perspective.

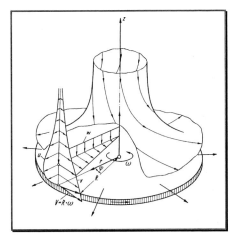

Figure 4.2. Flow close to a rotating disk in a fluid at rest

4.2.2   Analytical solution

Taking into account rotational symmetry as well as the notation for the problem, all derivatives in the circumferential direction are zero and no variables depend on the circumferential coordinate. We can rewrite the continuity Equation (2.7) and Navier-Stokes Equations (2.8-10) in cylindrical coordinates as

$$\frac{\partial U}{\partial r}+\frac{\partial W}{\partial z}+\frac{U}{r}=0, \qquad (4.1)$$

$$U\frac{\partial U}{\partial r}-\frac{V^2}{r}+W\frac{\partial U}{\partial z}=-\frac{1}{\rho}\frac{\partial P}{\partial r}+\nu\left[\frac{\partial^2 U_r}{\partial r^2}+\frac{\partial}{\partial r}\left(\frac{U_r}{r}\right)+\frac{\partial^2 U_r}{\partial z^2}\right], \qquad (4.2)$$

$$U\frac{\partial V}{\partial r}+\frac{UV}{r}+W\frac{\partial V}{\partial z}=v\left[\frac{\partial^2 U}{\partial r^2}+\frac{\partial}{\partial r}\left(\frac{V}{r}\right)+\frac{\partial^2 V}{\partial z^2}\right], \tag{4.3}$$

$$U\frac{\partial W}{\partial r}+W\frac{\partial W}{\partial z}=-\frac{1}{\rho}\frac{\partial P}{\partial z}+v\left[\frac{\partial^2 W}{\partial r^2}+\frac{1}{r}\frac{\partial W}{\partial r}+\frac{\partial^2 W}{\partial z^2}\right]. \tag{4.4}$$

Here the dynamic viscosity μ is replaced with the kinematic viscosity v with the relation $v=\mu/\rho$. For a disk with an infinite diameter, according to the no-slip condition at the wall, the following boundary conditions can be given:

$$\begin{array}{llll} z=0: & U=0, & V=\omega r, & W=0, \\ z=\infty: & U=0, & V=0. \end{array} \tag{4.5}$$

The layer thickness δ, which is carried by the disk, can be estimated in the following way. The centrifugal force per unit volume which acts on a fluid particle in the rotating layer at a distance r from the axis is equal to $\rho r\omega^2$. Hence for a volume of area $drds$ and height δ, the centrifugal force is $\rho r\omega^2 \delta drds$. The centrifugal force is balanced by a shearing stress τ_w, which acts upon the same element of fluid, pointing in the direction in which the fluid is slipping, and forming an angle θ with the circumferential velocity. It satisfies

$$\tau_\omega \sin\theta drds = \rho r\omega^2 \delta drds \quad\Rightarrow\quad \tau_\omega \sin\theta = \rho r\omega^2 \delta. \tag{4.6}$$

On the other hand the circumferential component of the shearing tress must be proportional to the velocity gradient of the circumferential velocity at the wall. This condition gives

$$\tau_\omega \cos\theta \propto \mu r\omega/\delta. \tag{4.7}$$

Eliminating τ_ω from these two equations, we obtain:

$$\delta^2 \propto \frac{v}{\omega}\tan\theta. \tag{4.8}$$

Assuming that the direction of slip in the flow near the wall is independent of the radius, the thickness of the layer carried by the disk becomes

$$\delta \propto \sqrt{\frac{v}{\omega}}. \tag{4.9}$$

In order to integrate the system of equations, a dimensionless distance from the wall can be introduced as

$$\zeta = z\sqrt{\frac{\omega}{\upsilon}}. \qquad (4.10)$$

Further, the following assumptions are made for the velocity components and the pressure:

$$U = r\omega F(\zeta); \qquad V = r\omega G(\zeta); \qquad W = \sqrt{\upsilon\omega}H(\zeta),$$
$$P = P(z) = \rho\upsilon\omega P(\zeta). \qquad (4.11)$$

Inserting these equations into Equations (4.1)-(4.4), a system of four simultaneous ordinary differential equations for the functions F, G, H and P can be obtained:

$$2F + H' = 0,$$
$$F^2 + F'H - G^2 - F'' = 0,$$
$$2FG + HG' - G'' = 0, \qquad (4.12)$$
$$P' + HH' - H'' = 0.$$

The boundary conditions in Equation (4.5) can be calculated as

$$\zeta = 0: \quad F = 0, \ G = 1, \ H = 0, \ P = 0,$$
$$\zeta = \infty: \quad F = 0, \ G = 0. \qquad (4.13)$$

With the help of similarity solutions, the system of Navier-Stokes equations is simplified to a system of ordinary differential equations. A fourth-order Runge-Kutta integrator with a Newton-Raphson searching method was applied to solve these ordinary differential equations. As an alternative, a finite volume method applying the tridiagonal-matrix algorithm (TDMA) to solve the algebraic equations suggested by Patankar (1980) [93] was also employed to solve this system. Good agreement between the results of the two methods was achieved. Both methods are introduced in detail in Appendices A and B. The results are plotted in Figure 4.3.

It emerges that the circumferential velocity has dropped to within 1% of the disk velocity at $\xi = 5.5$. The thickness of the boundary layer is

$$\delta = 5.5\sqrt{\frac{\upsilon}{\omega}}. \qquad (4.14)$$

The inclination of the relative streamlines near the wall with respect to the circumferential direction can be calculated by

$$\tan\varphi_0 = -\left(\frac{\partial U/\partial z}{\partial V/\partial z}\right)_{z=0} = -\frac{F'(0)}{G'(0)} = \frac{0.510}{0.616} = 0.828. \qquad (4.15)$$

Therefore, the angle is $\varphi_0 = 39.6°$.

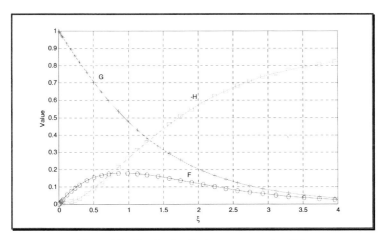

Figure 4.3: Velocity distribution near a rotating disk

Note that the results obtained from the similarity solutions of an infinite disk can be applied to a circular disk of finite radius R provided that the radius R is large compared with the thickness δ of the layer.

4.3   Numerical Investigation of Rotating Disk Flow

4.3.1   Introduction

In order to study the instabilities in the rotating flow, the flow generated by a rotating disk in a square tank numerically was investigated. The numerical calculations with different values of Reynolds number were carried out both for the steady averaged flow in which all time-dependent terms were neglected and for the time-dependent flow where the time-dependent Navier-Stokes equations were solved.

To reduce the computation cost, for steady flows the numerical simulation was carried out in a quarter of the domain under the assumption that a 4-fold periodicity of the solutions exists owing to the symmetry of the geometry. In this case a finer grid distribution can be achieved within the same level of the computer time. For unsteady flows, since the instabilities in rotational flows are normally associated with the asymmetry of flow field pattern, the numerical calculations were done in the full geometry arrangements. The simulated results were compared with the analytical similarity solutions described in the last section. Based on these comparisons, the grid level was determined for unsteady calculations. An imaginary lid was employed in the numerical investigations, simplifying the free surface treatment. The detailed dimensions of the computational domain are shown in Figure 4.1.

4.3.2   Grid structure and level

The disk rotation makes the system possess localized flow features where the
gradient of the velocity is very large in
the vicinity of the disk. In order to illus-
trate the local properties of the flow, especially the flow near the disk, the domain was decomposed into several separate blocks with different grid resolutions.
A finer grid was applied in the blocks
near the disk. Also, the grid density in
individual blocks was non-uniform, the
grid density being set higher at the sides
closer to the disk. Figure 4.4 illustrates
the block structure schematically.

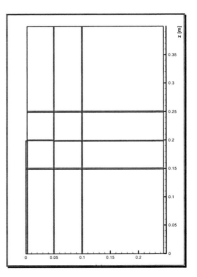

Figure 4.4: Schematic presentation of the used
block structured grid

In order to study the grid level sensitivity,
a very fine grid for the quarter geometry
was chosen in order to obtain the most
accurate simulation results within the capacity and performance of the computer employed.
For the full geometry simulations, three different numerical grid levels were applied. The control volume numbers used in the corresponding calculations are summarized in Table 4.1.

Grid level	Full geometry	Quarter geometry
Fine	601,600	644,352
Medium	198,912	-
Coarse	86,912	-

Table 4.1: Number of control volumes (CVs)

4.3.3   Results and discussion

4.3.3.1   Steady flow calculations

For steady flow calculations, the rotational speed was set to 10 rpm, and hence a Reynolds
number of the rotating disk $Re = \omega D^2/\nu = 1667$, ensuring flow in the laminar regime.

Large-scale flow field

Figure 4.5 shows the predicted flow field in a square tank with a fine grid level in the case of
full geometry for both vertical and horizontal cross-sections. For representation of the results,

the spatial coordinates x, y and z were normalized by the width T of the tank and velocities by the tip velocity of the disk V_{tip}. The predicted results with the other grid levels have a similar flow field pattern, but, differ in details such as locations of the loop centre point, maxima of velocities, etc., with respect to the grid levels used. In Figure 4.5-a the flow field in a vertical cross-section parallel to the tank wall is presented. The circumferential component V is represented by the contours labelled by the contour legend.

In the plane of the disk, the fluid is thrown out by the centrifugal forces generated by the rotation of the disk, forming a radial jet flow outwards to the tank wall. The radial velocity increases with the distance from the disk up to a maximum of about 0.075 V_{tip} at about 0.168 D from the edge of the disk. This radial jet has an average thickness of about twice the disk thickness at the very beginning and grows gradually owing to the "entrainment" of the fluid from the surrounding region into the jet. In the radial jet flow there is almost no axial velocity component. Directly in the vicinity of the rotating disk a much higher circumferential velocity component with a maximum magnitude of about 0.15 V_{tip} appears, whereas in other regions this component is 10-100 times smaller. The radial jet flow hits the wall and splits into upward and downward flows. Owing to the continuity, in the region both above and below the disk the fluid is drawn back into the disk region, hence two almost symmetrical circulation loops are formed.

Figure 4.5-b presents the flow pattern in three horizontal planes, namely in the disk plane and in both middle planes positioned in the lower and upper parts separated by the disk. In the disk plane, the fluid leaves the rotating disk in the tangential direction with a velocity almost the same as the tip velocity of the disk, and decreases with the distance from the disk very quickly. The inclination angle between the velocity and the circumferential velocity φ varied from nearly 0° at the disk edge to about 25.6° in the direct vicinity, and increased to 73.6° shortly before reaching the side wall. On the top of disk this angle remained in the range around 31.8°, a small deviation from that (39.6°) of the similarity solutions as in Section 4.2.2. The flow pattern in the lower and upper middle plane showed considerable similarity, except in the vicinity of the rotation axis where there is a shaft in the upper part, distinguishing these two parts. The flow pattern stayed fairly axisymmetric, except in the region near the side wall. Owing to the square shape of the tank, the flow was drawn into the centre after fluid passing the diagonal plane of the tank along the direction of rotation more strongly, indicating that the square tank arrangement has a similar obstacle effect to baffles in cylindrical vessels. The 4-fold inward spirals are clearly to be seen in the range corresponding to the diameter of the disk. Near the central point the flow is evidently axisymmetric.

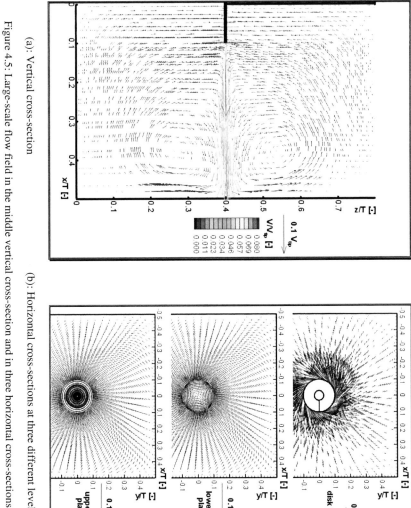

(a): Vertical cross-section

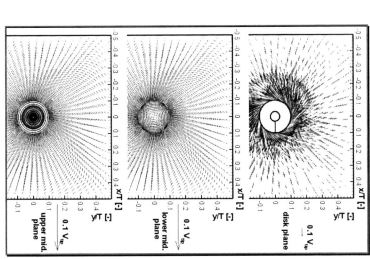

(b): Horizontal cross-sections at three different levels

Figure 4.5: Large-scale flow field in the middle vertical cross-section and in three horizontal cross-sections

Von Kármán boundary layer

The accuracy of the predicted flow field above and below the rotating disk can be better estimated by comparing them with the similarity solutions of the Von Kármán layer discussed in
Section 4.2.2. Here, for convenience, the spatial coordinates were set to cylindrical coordinates and normalized with the disk radius R. Vertical profiles at $r/R = 0.5$ both below and
above the disk were compared with the theoretical solutions.

The comparison confirmed the disk symmetry of the flow field except in the vicinity of the
shaft, since all three components have identical magnitudes between the flow above and below the disk. Figure 4.6-a and b show the comparative results for the radial and tangential
components. The velocities present good agreement between the numerical and analytical
solutions. The small deviations near the disk surface can be attributed to the discretization of
the numerical grid. In comparison, the predicted axial component departs from the analytical
solution considerably both in profile shape and in magnitude, as depicted in Figure 4.6-c.
There are two possible reasons for these deviations. One is that the computation was carried
out in a limit domain and zero velocity was imposed at the limited boundary, while the velocity of Von Kármán boundary layer solutions tends to zero at infinite height. Figure 4.6-c
shows that the simulated axial velocity tends to zero more quickly. As the second reason, the
Von Kármán boundary layer solution is obtained under the assumption that both the radial
and tangential components are proportional to the radial distance from the rotating centre;

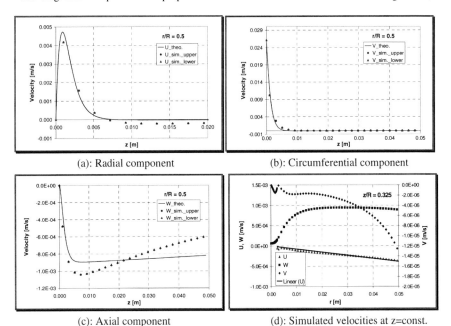

(a): Radial component   (b): Circumferential component

(c): Axial component   (d): Simulated velocities at z=const.

Figure 4.6: Comparisons between simulated and theoretical results near the Von Kármán layer

however, this assumption was not satisfied in the numerical solution. Figure 4.6-d illustrates the contradiction of this assumption in the numerical calculation. In comparison, the radial component agrees well with the linearity assumption, whereas for the tangential component, except that in the region very close to the centre point, linearity exists. In most regions the tangential component showed a paraboloic shape.

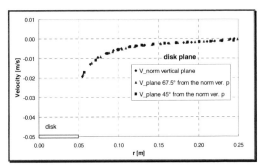

Figure 4.7: Comparison of tangential components in different vertical planes

Another assumption in the theoretical solutions, that all values are axisymmetrical, was confirmed by Figure 4.7 despite the non-axisymmetry of the square tank geometry. As an example, the tangential components in three different vertical planes were compared in the norm cross-section, plane 45° and 67.5° from the norm cross-section. The acceptable deviations between the three planes are attributed again to the square geometry of the tank. The other two components show better agreement. This confirms quantitatively the axisymmetry conclusion drawn from Figure 4.5-b.

Radial jet flow in the disk plane

The accuracy of the numerical solutions in the bulk flow, which is away from the disk boundary layer, can be also estimated. By virtue of the flow symmetry about the disk plane, the axial velocity component satisfies

$$W = 0 , \frac{\partial W}{\partial z} = 0 . \tag{4.16}$$

Moreover, owing to the axisymmetry, $\frac{\partial}{\partial \phi}() = 0$. The continuity Equation (2.7) can be simplified by

$$\frac{\partial U}{\partial r} + \frac{U}{r} = 0 \Rightarrow \frac{1}{r}(\frac{\partial (U \cdot r)}{\partial r} - U) + \frac{U}{r} = 0 . \tag{4.17}$$

It follows that

$$\frac{\partial (U \cdot r)}{\partial r} = 0 \Rightarrow U = \text{constant} / r \tag{4.18}$$

The radial profile in the disk plane should satisfy the curve U=constant/r.

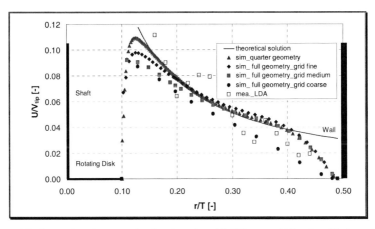

Figure 4.8: Comparison between simulated results with different grid levels, with theoretical and LDA measurement results

Figure 4.8 presents a comparison of the analytical solution of Equation (4.18) with the numerical solution with three different grid levels. The radial velocity component U in the disk plane is normalized by V_{tip}. In addition, the radial velocity component in the disk plane was also measured experimentally by LDA. However, the LDA measurements were carried out at a higher rotational speed (300 rpm) since the low rotational speed (10 rpm) that was applied in the numerical investigation was not stable with respect to the motor and the controlling units in the experiments. Moreover, at a very low rotational speed, it was also difficult to obtain an acceptable LDA measuring rate. For comparisons, the LDA measurement results were also normalized by the tip velocity of the rotating disk.

From the comparison, it can be concluded that for the quarter geometry and the full geometry with the highest grid level, the simulated results agree well with those of the theoretical solution (Equation (4.18)), but only in the medium range. Near the wall region, there are relatively large deviations due to the presence of the wall in the numerical solution, while the influence of the wall was not taken into account in Equation (4.18). In the vicinity of the rotating disk, a large velocity gradient exists, the velocity increasing sharply with increasing distance from the disk edge, and achieving a maximum at about $r/T = 0.125$ (corresponding to 0.168 D). After reaching the maximum, the velocity decreases along the theoretical profile up to ca. $r/T = 0.4$. Therefore, in the region where a large velocity gradient exists, a finer grid is needed to follow the velocity variations more accurately. Figure 4.8 shows that the numerical solution tends to the analytical solution as the grid step tends to zero.

The LDA measurements show a similar profile shape. However, larger deviations fluctuating around the theoretical solutions can easily be seen. This might be attributed, on the one hand, to the fact that at a higher rotational speed the flow becomes turbulent. On the other hand, instabilities were involved into the measurement results. The clear Von Kármán vortex street appearing in the radial jet flow and also the precessing motion of the rotating centre shown in the visualization, which will be described further in Section 4.4, might be the main reason for these deviations.

4.3.3.2   Unsteady flow calculations

Lower rotational speed

In order to have an overview of the flow developing from the beginning of the disk rotation, the calculation was started from the instant at which the disk began to rotate. The rotational speed of the disk was 10 rpm, and the computation was carried out with the medium grid in the whole domain. The time step was set to 0.01 s, thus with a Courant number $Co = 0.77 < 1$, ensuring convergence in each time iteration. The flow field calculated from the beginning of the disk rotation is illustrated in Figure 4.9. Since all velocity components at the beginning of the rotation are nearly zero except in the region near the disk, the flow field was represented in a uniform magnitude in which only the velocity directions were shown.

At $t = 0.02$ s, the flow motion spreads out in all directions from the disk in a spherical form. At the same time, the fluid in other regions begins to move, but with a very small magnitude (nearly zero). Owing to the conservation of mass, the fluid is drawn back into the disk region. The rotation of the shaft also has an influence, as shown in the middle figure ($t = 0.08$ s). After about 0.2 s, two well formed circulation loops can be easily seen in the right figure. After this instant the flow does not change its main pattern and increases mainly in magnitude. This developing process is clearly illustrated in Figure 4.10. The profiles of the radial and tangential velocity components in the middle plane of the upper tank ($z = 0.3$ m) were compared at different instants. At the beginning, both radial and axial components are nearly

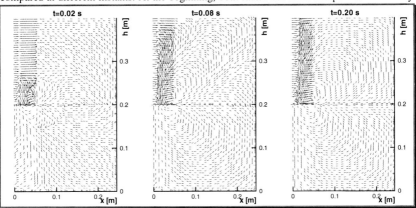

Figure 4.9: Time series of the rotating disk flow at n=10 rpm

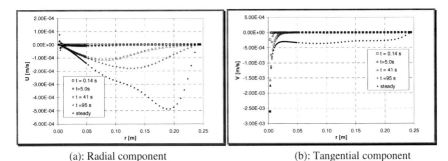

(a): Radial component   (b): Tangential component

Figure 4.10: Comparison of velocity components at different times

different instants. At the beginning, both radial and axial components are nearly zero, then the tangential component increases more quickly and subsequently, after about 40 s, very slowly. Close to the edge of the rotating disk, the tangential component decreases sharply and tends to zero with increasing radial distance from the axis. As time proceeds, the non-zero part extends radially outwards. It takes about 250 s until the velocities approach the steady state.

After the steady-state situation of the fully developed flow, the computation of the unsteady flow started with fully developed steady flow as the initial conditions. Full geometry with a fine grid level was selected for this calculations. At each time step the calculation had at least 1000 iterations. Attention was paid to the lower part of the tank, where a better overview of the flow developing process can be obtained without the perturbation of the shaft. Figure 4.11 presents a comparison of the flow field below the rotating disk ($z = 0.1$ m) at two different instants. The flow pattern at any two time steps is approximately identical with that of the steady state. No remarkable fluctuations and instability occur. The flow in the tank is steady.

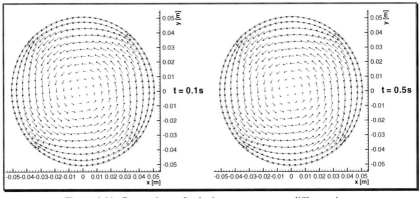

Figure 4.11: Comparison of velocity components at different times

Higher rotational speed

As the rotational speed of the rotating disk was increased from 10 to 100 rpm, corresponding to a disk Reynolds number $Re = 16,667$, the standard k-ε turbulence model was integrated into the calculations.

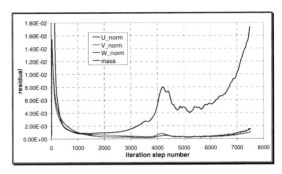

Figure 4.12: Residual history of the calculations at higher rotational speed

The most remarkable result is that even the steady flow calculations tended to be unstable. This unsteadiness can be reflected by the development of the residual of variables involved in the calculations. Whereas in the calcula-

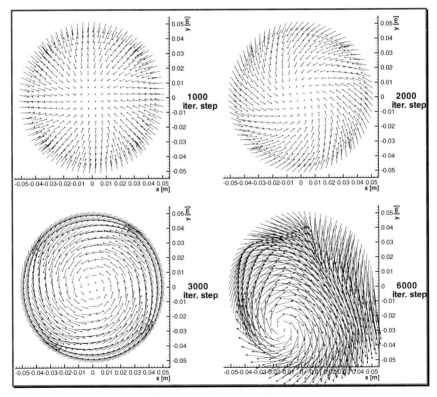

Figure 4.13: Different flow patterns at different iteration steps

tions with a lower rotational speed the residual used for the determination of convergence decreased very quickly with the iteration number, in the calculation with a higher rotational speed, after a certain number of iterations, the residuals fluctuated and ultimately increased sharply, as shown in Figure 4.12. All residuals at first decreased quickly, after about 1500 iterations the residual of the mass tended to increase with the iteration steps, and after 3000 iterations the mass residual began to jump suddenly. After a small backward fluctuation between about 4100 and 5200 steps, the residual increased strongly and the calculations ultimately diverged. The required convergence criteria could no longer be achieved. In most cases, this phenomenon appears in the calculations as a symptom of the instability of the flow considered [72]. More interesting is the flow pattern calculated after different iteration steps presented in Figure 4.13. After 1000 iteration steps, strong radial inward flow to the rotating centre point can be clearly seen. The flow pattern is totally different to that at a lower rotational speed. An 8-arm spiral inward flow developed after 2000 iteration steps with a wave front angle $\varepsilon \approx 15°$, indicating the existence of an SRI instability as mentioned in Section 2.1.7.2. In the computation, after 3000 iteration steps the tangential component increased in intensity, and an approximate axisymmetric spiral flow developed. After the residual jumped strongly after 6000 iteration steps, the flow became totally asymmetric. The rotating centre point departed from the geometry centre point. This departure of the geometric rotating centre point corresponds to the PVC phenomenon mentioned in the last section.

4.4 Experimental Investigation of Rotating Disk Flows

4.4.1 Laser sheet visualization in rotating disk flows in a square tank

The flow pattern and the associated instabilities for the rotating disk flow in a square tank were also investigated experimentally by the laser sheet visualization technique. The experimental set-up is depicted in Figure 4.1 and was identical with the geometry in the numerical investigations, but with a free surface on the top of the liquid content. For further comparisons, a second disk with double the diameter ($D = 200$ mm) was investigated. An electric DC motor was used to drive the disks. Because of the difficulty in maintaining a stable angular velocity for low rotational speeds, no visualizations were taken for $N < 50$ rpm. With the relatively strong vibration of the shaft and disk in the resonance range (200-500 rpm), it was not the initial objective to make a transition diagram of the instabilities in the present work. The visualization results were recorded both photographically and by video. However, owing to the relatively long exposure time (about 1 s), it was difficult to capture the instability phenomena exactly. Therefore, the results were mainly analysed from the videos.

It can be seen from the laser light sheet visualizations that in addition to the primary circular motion, the rotating disk generates a secondary flow pattern as in the calculation results mentioned in the last section. Two circulation loops appear in the vertical cross-section, but, the location of both loop central points is not stable as in the calculations. Also, a remarkable feature is that the radial jet flow driven by the centrifugal forces leaves the rotating disk along a similar Von Kármán vortex street. This motion was not found in either steady or unsteady calculations. This deviation is attributed to the unavoidable geometric and mounting errors in the experimental investigation, these errors acting as disturbances in the radial jet flow. This radial jet flow splits into two directions at the wall, forming two circulation loops. The fluid is drawn back into the disk region upwards and downward in the region near the rotation axis.

The most interesting phenomenon is the motion of the vortex core. At low rotational speeds ($N < 80$ rpm), regardless of the vibration of the disk, the rotation flow centre, namely the vortex core, was retained in the geometry centre both in the upper part, where a shaft exists in the geometry centre, and in the lower part. As the rotational speed increases, the vortex core begins to depart from its original position, and is brought into a precessing motion. This precessing motion is depicted in Figure 4.14-a. The departure off-distance and the rotational speed of the precessing vortex core increase with the increasing rotational speed of the disk, as shown in Figure 4.14-b. Owing to the presence of the shaft and its rotation, only by very careful observations could the same precessing motion in the upper part above the disk be found. The vortex core is separated into two parts by the disk. Both parts rotate with the same speed, but not the rotational speed of the disk. At further higher rotational speeds, the precessing of the vortex core in the upper part of the tank appears in the form of a vertical vortex at the free surface, which is rotating around the shaft. The form of the vortex core itself departs from its direct line into a spatial spiral form. The precessing motion of the vortex core appears already at relatively low rotational speeds, as the disk diameter was increased to 200 mm. The depar-

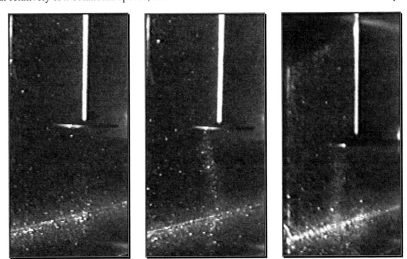

(a):$d = 100$ mm, $N = 200$ 1/min (b): $d = 100$ mm, $N = 600$ 1/min (c): $d = 200$ mm, $N = 300$ 1/min

Figure 4.14: Laser light visualization for rotating disk flows

ture off-distance also grows as shown in Figure 4.14-c.

It can be assumed that the vortex core in the rotating flow tends to depart from its geometry centre. Disturbances, such as baffles, tiny geometry imperfections, operating conditions, etc., can accelerate this tendency significantly.

4.4.2   Steady flow investigations in stirred vessels

4.4.2.1   Introduction and experimental set-up

The rotating disk flow was also measured by LDA in a standard full-baffled tank, and was compared with the stirred flow in the same vessel induced by a standard radial impeller, a six-blade Rushton turbine. The tank had a diameter of 152 mm, and detailed dimensions are listed in Table 3.1. The thickness of the disk was 2 mm and the diameter of the disk was 50 mm ($D/T = 0.33$), simplifying quantitative comparison with the RT, which had the same diameter. The LDA measurements for the rotating disk and for the RT were carried out previously by Kunte (1996) [70] and Schäfer *et al.* (1995) [107]. The flow field for both cases was measured at the same Reynolds number ($Re = 1200$). Table 4.2 lists the dimensions of the RT. More details can be found elsewhere [70, 109].

D	D_{Disk}	D_{Hub}	D_{Shaft}	L_{Blade}	W_{Blade}	T_{Blade}
50 mm	37.5 mm	12.5 mm	8 mm	12.5 mm	10 mm	1.75 mm
(0.33 T)	(0.247 T)	(0.082 T)	(0.053 T)	(0.082 T)	(0.066 T)	(0.011 T)

Table 4.2: Geometric dimensions of RT

4.4.2.2   Results and discussion

The flow fields in both cases were measured in different vertical planes depending on the baffle locations. Here, only the flow fields in the middle plane between two neighbouring baffles are presented in Figure 4.15-a for the disk and Figure 4.15-b for the RT. All velocities were normalized by the tip velocity of the disk and the impeller blade.

The most remarkable feature in the comparison is the close similarity of the circulation loops of the flow fields. In both cases, two large-scale circulation loops exist, separated in the rotating disk plane or the stirrer plane. This is also in good agreement with the flow pattern in the numerical investigations for the rotating disk flow but in a square tank. The centre point of the circulation loops in both cases deviates slightly in position. For the disk, for instance, the central point of the upper loop is located at point $r/T = 0.316$, $z/T = 0.461$, whereas for the RT it is at $r/T = 0.382$, $z/T = 0.5$. Hence the loop central point for the RT is pushed slightly outwards. This difference can be attributed to the much stronger radial jet flow in the stirrer plane of the RT. Whereas the radial jet flow for the disk is ejected in the purely radial direction, strong irregular velocity directions, especially in the vicinity of the stirrer region, appear in the radial jet flow of the RT, indicating the existence of a trailing vortex structure generated by the stirrer blade [109]. The upper circulation loop in both cases does not reach the filling height; for the RT the upper loop achieves a height of ca. $z/T = 0.72$, and for the disk only ca. $z/T = 0.62$.

This secondary circulation, however, differs strongly in intensity. The maximum of the radial velocity component for the disk appears exactly in the disk plane with a magnitude of 0.02 V_{tip}, whereas for the RT up to 0.85 V_{tip} is achieved in the stirrer plane. The magnitudes of all velocity components for the disk in the whole tank are much smaller than those of the RT (of the order of about 0.01 of the RT).

From the above comparisons, it can be seen that the secondary flow fields of a rotating disk and of a standard radial impeller, e.g. an RT, are very similar in circulation pattern except in the impeller region, where for an RT trailing vortices are generated. However, the velocities differ strongly in magnitude. The vertical mounted stirrer blades of the RT intensify both the primary motion and the secondary circulations. Therefore, the stirred flows possess the basic features of the rotational flows, e.g. the precessing of the vortex core. The stirred flows will be discussed in Chapter 6 further.

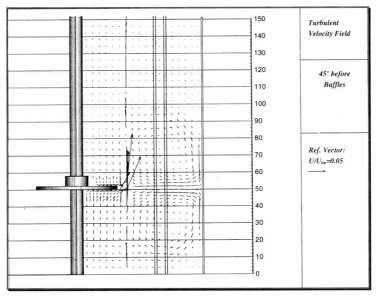

(a): Large-scale flow field of the rotating disk in the middle plane between baffles

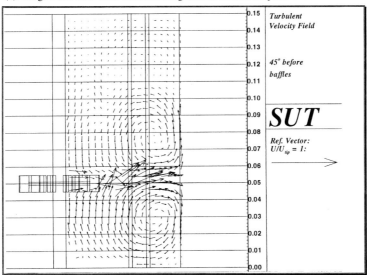

(b): Large-scale flow field of the RT in the middle plane between baffles

Figure 4.15: Comparisons of flow patterns between the disk and the RT

5 PRIMARY RESULTS OF VISUAL OBSERVATIONS ON SURFACE AERATION

The mechanisms of surface aeration as reported by numerous research groups are essentially different descriptions of the same phenomenon. It was widely accepted that various types of surface vortices, waves and high turbulence level at the surface are responsible for air entrainment. However, an adequate description of the explanation and the features of the formation of surface vortices have seldom been presented. In this chapter, the formation and the primary features for two typical impellers, a radial Rushton turbine (RT) and an axial pitched blade turbine (PBT), are described based on the visual observations made in the present work. The visual criteria to determine the critical impeller speeds for the onset of surface aeration are refined and their accuracies are compared. Based on the results of visual observation, physical modelling and dimensional analysis are employed to estimate the empirical correlations existing in the literature and to select appropriate correlations describing the phenomenon with a more physical background.

5.1 General mechanisms of surface aeration in stirred vessels

5.1.1 Unbaffled vessels

The flow in an unbaffled dished-bottomed vessel of diameter $T = 350$ mm was observed. The impeller was a standard six-blade RT having a diameter $D = 132$ mm ($D/T = 0.38$). The bottom clearance C was set at $C/T = 0.33$. The tank was filled with water ($H/T = 1$) and had a free surface.

The liquid rotates together with the stirrer in the unbaffled vessel. Near the centre of the vessel, the liquid rotates with the same angular velocity close to that of the impeller. This motion is the so-called solid body rotation (forced vortex). Little mixing takes place as a consequence of the absence of velocity gradients. The flow in the outer part is similar to that of a free vortex. Hence, the whole flow in unbaffled tanks is a combination of two types of vortices, i.e. the Rankie vortex.

With increase in the impeller rotational speed, a large surface central vortex forms in the centre of the rotating fluid owing to the centrifugal forces of the impeller throwing the fluid out to the walls. The depth of this central surface vortex depends directly on the rotational speed of the stirrer and thus the rotation of the fluid. The structure of the central vortex is shown in Figure 5.1-a. The surface shape is, in general, in good agreement with the analytical pressure distribution of a Rankie vortex shown in Figure 2.4-b. The correlation between the maximum impinging depth of the free surface and the rotational speed of the stirrers has been well investigated [7,152]. In their correlations, the critical parameter was the Froude number Fr. A simplified potential flow theory of the vortex geometry in unbaffled stirred vessels was also

 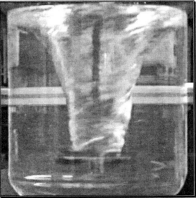

(a): Central vortex profile   (b): Vortex depth up to stirrer

Figure 5.1: Formation of the central vortex in unbaffled vessels

investigated by Nagata (1975) [89] theoretically and verified with his own measurements. More recently the surface structure was also studied numerically by Ciofaloet *et al.* (1996) [24]. In all these investigations, it was assumed that the profile of the central vortex has a Rankie vortex structure. At the centre of the tank the fluid is in a forced vortex motion. Outside this region the free vortex motion is dominant. The transition point of the two vortex regions was calculated between $r/R = 0.60$ and 0.65 for stirrers with D/T ratios 0.3-0.7. This value was estimated by visual observation in the present work as 0.61 ($r \approx 40$ mm), which agrees well with the work of Nagata [89].

With higher rotational impeller speeds, the secondary flow pattern increases in intensity. The downwards axial velocity is superimposed on the rotational flow in the centre part. The spatial spiral swirling flow is intensified. This is symbolized by the spiral stripes surrounding the air-liquid vortex phase boundary shown in Figure 5.1-b. It holds its regular form except at the bottom of the vortex where the vortex boundary changes its shape in an irregularly leaping way. During the development of the central vortex in depth and intensity, the phase boundary is well defined and no air bubbles can be entrained into the liquid phase. As soon as the bottom reaches the stirrer region with increasing impeller speed, air pockets are brought from the air phase into the stirrer region, where they are broken into small bubbles and ultimately dispersed into the liquid phase. This type of air entrainment is also illustrated in Figure 5.1-b. The central vortex is the exclusive mechanism of surface aeration in unbaffled stirred vessels.

The central vortex in unbaffled vessels can be suppressed by applying baffles effectively. Some other methods can also be helpful in suppressing the occurrence of the central vortex, such as a square tank geometry, eccentric or inclined mounting of the stirrer, etc [125].

5.1.2   Standard baffled vessels

5.1.2.1   Experimental set-up

Most of the visual observations were carried out in the same tank as used for LDA measurements. It has a flat bottom and a diameter of $T = 400$ mm. Four equi-spaced vertical baffles were mounted in the tank. The detailed dimensions of the tank and the baffles can be found in Table 3.1. The visual observations were carried out both in water and in a silicone oil mixture which was used for LDA measurements. For investigations of the mechanism of surface aeration, only the results obtained in water are presented here. The filling height was 400 mm ($H/T = 1$) with a free surface. Two standard stirrer elements, a radial Rushton turbine (RT) and an axial pitched blade turbine (PBT), were employed in the visual observations to cover the variety of impeller types. The dimensions are listed in Table 1.1.

Stirrer type	D	D_{Disk}	D_{Hub}	D_{Shaft}	L_{Blade}	W_{Blade}	T_{Blade}
RT	132 mm	99 mm	29.5 mm	20 mm	33 mm	26.4 mm	5 mm
	(0.33 T)	(0.248 T)	(0.074 T)	(0.05 T)	(0.083 T)	(0.066 T)	(0.013 T)
PBT	132 mm	-	36.8 mm	20 mm	47.6 mm	26.4 mm	2.5 mm
	(0.33 T)		(0.092 T)	(0.05 T)	(0.12 T)	(0.066 T)	(0.006 T)

Table 5.1: Geometric dimensions of RT and PBT

The rotational speed of the stirrer was controlled by adjusting the supply voltage of a DC-motor mounted on the top of the shaft. The temperature of the working fluid was kept constant by a cooling system fixed at the vessel bottom, in order to filter out the effect of temperature on the liquid viscosity and surface tension. The observation results were recorded with a Panasonic video camera and ultimately stored in a PC for further analysis.

5.1.2.2   General mechanisms of surface aeration

In the lower range of the impeller speed, the free surface remains calm and very low mixing activity takes place. The fluids near the surface act like a solid motion. There is no significant relative motion between liquid sheets (see Figure 5.2-a). This status implied the steeping flow regime mentioned in Section 2.2.4, where the influence of baffles is negligible. With increasing rotational speed, first, small dimples appear and they gradually form weak waves. The surface does not retain its strictly horizontal level any longer.

As the stirrer rotational speed increases continuously, the surface becomes more active. The influence of the baffles is obviously visible, especially in the region near the free surface. The flow is redirected from a pure circumferential motion into a complicated radial-tangential flow. Different forms of swirl vortices appear at the surface whose axis is perpendicular to the surface plane (Figure 5.2-b). These vortices are normally called "surface vortices" or "macro vortices" by Liepe *et al.* [73], which will be discussed further in the next section. These vortices are driven and disturbed by the large-scale stirred flow at the surface. These surface vortices are unsteady, and vary in both shape and intensity with time and place. In the range of surface vortices, the fluid has to follow the bulk flow together with the vortices and at the same time rotates around the surface vortex axis in a spiral form. The angular velocity near the vortex core has a significant magnitude compared with that of the surrounding fluid, and also possesses a strong downwards component. The surface of these vortices has a similar shape to a concentrated Rankie vortex, but the core of the forced vortex part is very small.

With further increasing in the impeller rotational speed, these vortices grow in both intensity and unsteadiness. The disturbances from the large-scale stirred flow, especially redirected flow by baffles, are strong enough to lead to impinging and enfolding of the phase boundary between air and the liquid phase at the bottom of these surface vortices. Single air bubbles can be enfolded at the bottom of surface vortices (Figure 5.2-c). On increasing the rotational speed continuously, the disturbances introduce more bubbles in the same way and these bubbles accumulated in cloudy bubble clusters (Figure 5.2-d). They are forced to move with vortices like comet tails. In this process larger bubbles are broken into many smaller ones by the high shear stress in the centre of the vortices and impinging between bubbles. As the rotational speed is increased further to a certain level, the downward velocities in the bulk circulation loops generated by the stirrer are great enough to overcome the bubble rise velocity (Figure 5.2-e). In this state the bubbles can be transported into the stirrer region and be broken into still smaller bubbles there owing to the high shear stress in the region. Ultimately they are dispersed into the whole vessel following the bulk flow circulation.

With still higher rotational speeds, air bubbles can be introduced continuously and a steady surface aeration state is thus achieved (Figure 5.2-f). In addition, the free surface up to this state becomes very unsteady and wavy. Larger air bubbles can also be entrapped in the liquid by wave impinging, contributing to the intensity of surface aeration. The gas hold-up and the mass transfer between the two phases reach a certain level. Bubbles are massively entrained into the vessel and form air cavities in the under-pressure region behind the stirrer blades. The average density in the impeller region is reduced, leading to a reduction in power consumption which is reflected by the sudden drop of the power number *Po* of the stirrer.

Generally, the initial process of surface aeration can be summarized in two steps. The first step is the entrapment of gas at the liquid surface due to high degree of disturbance and unsteadiness. The second step is the carriage of these bubbles into the tank from the liquid surface. The ability of the impeller to generate sufficient disturbance, which is often realized by the surface vortices and waves, and the ability to generate a favourable liquid phase flow pattern for efficient dispersion of the entrapped bubbles are the important parameters governing the phenomenon of surface aeration.

(a): Calm surface

(b): Surface with vortices

(c): Single bubbles

(d): Cloudy bubble clusters

(e): Separated bubbles from vortices

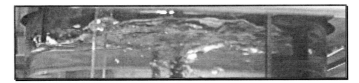

(f): Steady surface aeration

Figure 5.2: Visualization of the free surface in different rotational speed ranges

5.1.2.3   Other mechanisms of surface aeration

The intensity of the surface vortices increases strongly with decrease in the submergence depth of the stirrer in the liquid. This can be realized by either decreasing the liquid filling height *H* or increasing the stirrer bottom clearance *C*. In some situations, before the intensity of the surface vortices reaches the level to entrap air bubbles directly, a direct connection between the vortex bottom and the under-pressure region in the stirrer region can appear which is very similar to the "Taifun" phenomenon or "bathtub vortex". This "short-circuit" mechanisms is illustrated in Figure 5.3-a. The vortex can be visualized clearly by the concentrated air core. This air core is again broken into small air bubbles in the stirrer region. In this way air bubbles are entrapped much earlier in stirred vessels than in normal stirring configurations.

In the investigations of Genenger *et al.* [50], a similar surface vortex was found for a down-pumping axial impeller PBT with an eccentric shaft installation but in an unbaffled stirred tank. This vortex was responsible for air entrapment. However, they found that an up-pumping PBT in such a configuration could suppress the appearance of vortices and air entrainment successfully. The reason was attributed to the opposite flow direction of the circulation pattern in the centre of the tank. Owing to the upward flow direction in the central region near the shaft, an upward pressure drop is formed, which suppresses the formation of a central vortex. In the present work, an up-pumping PBT was checked for normal baffled stirred flows to avoid surface aeration. The investigation shows that this manner is only valid for highly viscous fluids (ca. 300 mPas in [50]) with an eccentric shaft installation in unbaffled tanks. For low-viscosity fluids which were of more interest in the present work, a different mechanism was found to be responsible for air entrainment under up-pumping conditions in a baf-

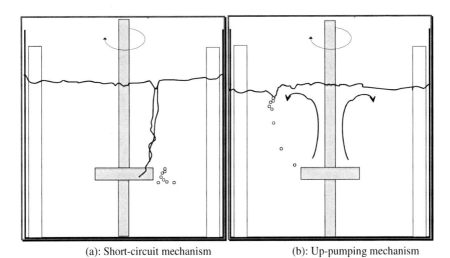

(a): Short-circuit mechanism   (b): Up-pumping mechanism

Figure 5.3: Schematic representation of other mechanisms of surface aeration in stirred tanks

fled tank. The surface vortices do not appear at the free surface. However, strongly upwards surges hit the surface in the region with dimensions comparable those of the impeller, and these surges emerge into the bulk flow again. At the rim of this region, air bubbles are entrained into the liquid owing to wave formation, accompanied by a high turbulence level. Figure 5.3-b represents this surface mechanism schematically.

5.1.3 Positions of appearance of surface vortices in baffled vessels

The positions of appearance of surface vortices are very dependent on the stirrer type and operating conditions, especially the former.

5.1.3.1 Radial stirrers

For radial stirrer RT, the surface vortices appear mainly in the region close to the stirrer shaft and rotate around the impeller shaft. Greaves and Kobbacy (1981) [52] were the first to consider the formation of this type of vortex. Figure 5.4 presents their observation results. As depicted, they assumed that a large stable cylindrical eddy A is formed from the flow generated by the impeller discharge interacting with the baffled tank wall. Thereafter, small eddies B and C are formed. They emphasized that eddies of type B have a tendency to rotate slowly around the impeller shaft and can form a hollow vortex. A similar rotating vertical vortex was also observed by Haam *et al.* (1992) [56]. Nevertheless, neither of these groups was able to explain the origin of this phenomenon.

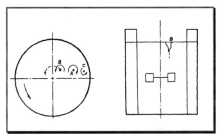

Figure 5.4: Schematic representation of vortex formation by Greaves and Kobbacy.

Generally, the flow in the central region of the tank has mainly a circumferential component. A strong circular motion of the fluid is again approached in spite of the presence of baffles. This can be attributed to the assumption of Clark and Vermeulen [25] that the influence of the baffles near the shaft is negligible. For a standard configuration with $H/T = 1$, one dominant vertical surface vortex appears close to the shaft, as shown already in Figure 5.2-b and c. More details about this dominant surface vortex are summarized in Figure 5.5. More often, the core of the dominant vortex moves along a nearly circular orbit as depicted in Figure 5.5-a. The radius of this orbit corresponds to the outer range of the solid body rotation region. Note that the motion of the surface vortex has its own sense and may not travel with the same velocity as the surrounding fluid. The motion of the vortex axis along the circular orbit behaves randomly. It can stay at one position for a short time and then sweep a stretch very quickly. Sometimes it moves also in a

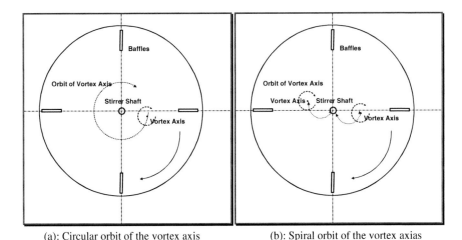

(a): Circular orbit of the vortex axis   (b): Spiral orbit of the vortex axias

Figure 5.5: Schematic representation of surface vortex formation and motion for RT

spiral trajectory towards the shaft and occasionally disappears, then suddenly runs out of the shaft again in a spiral way. This motion of the axis is described schematically in Figure 5.5-b. In addition, before each baffle surges swell at the free surface, which are the result of the re-bounded rotational flow by the baffles. The surges are further directed inwards to the central region by the baffles, and impinge on the rotating surface vortex. This additionally intensifies the unsteadiness of the surface vortex motion.

Statistically, if observation is carried out in a long period, the rotational motion of the surface vortex core has a stable rotational frequency, regardless of its instantaneous irregular behaviour. The rotational frequency of this vortex core f_v was estimated for the RT in the silicone oil mixture by counting the numbers of the vortex core passing a marked position in 5 min at different stirrer rotational speeds N. The counting was repeated at least 10 times to reduce the statistical deviations. The results are listed in Table 5.2.

N [1/min]	150	200	250	300
f_v [-]	0.11	0.13	0.18	0.21
f_v/N [-]	0.043	0.039	0.041	0.042

Table 5.2: Statistical rotational frequency of the surface vortex axis for RT in silicone oil

It can be seen from Table 5.2 that the rotational frequency of the vortex core increases propor-tionally with increasing rotational speed of the stirrers. Therefore, the rotational frequency of the vortex core f_v can be normalized by the impeller speed N. A constant dimensionless rotational frequency of the vortex core can be obtained, the value f_v/N being about 0.04 .

5.1.3.2   Axial stirrers

For PBT and other axial down-pumping stirrer elements, the surface vortices appear more irregularly in time and space compared with radial stirrer elements. In previous experimental and numerical investigations, the flow field was widely believed to be mainly a rotating flow and periodically steady in the whole tank, as soon as the flow achieved a steady state after certain stirrer rotations. This assumption contradicts the results of visual observations for the flow far from the stirrer, e.g. at the surface.

The flow in the region near the liquid surface was observed to be neither rotational nor periodically steady. The velocities are very low and the flow direction is not well defined. Occasional swells of the surface in the upstream area near the baffles appear. Before each baffle, strong upwards surges appear with an irregular rhythm, and the intensities of these surges before each baffle do not have the same level. As a consequence, the flow possesses a dominant direction, i.e., from the side with high intensity of surges to the lower side. Figure 5.6 illustrates the formation of this dominant flow direction. Behind the baffles downstream tail vortices are formed, these vortices being responsible for air entrainment in flows induced by axial impellers. Similar results were also obtained by Myers *et al.* (1997) [87] with digital particle image velocimetry (PIV) techniques.

In addition, this dominant flow direction varies with time since the surge before each baffle changes in intensity. The flow direction can be directed radially inward, and sometimes radially outward. During the changing of the flow directions, the fluid flowing in the new gen-

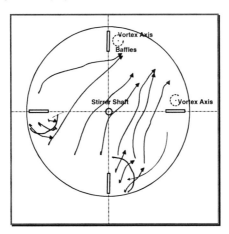

Figure 5.6: Schematic representation of the dominant flow direction at surface for PBT

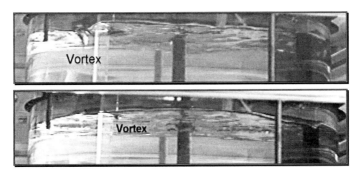

Figure 5.7: Surface vortices behind and before baffles for PBT

erated flow impinges on that of the last dominant direction. This impingement also contributes to air entrainment. The change of the dominant flow direction is random and the time that
a dominant direction lasts is also very irregular, from several up to tens of seconds. A statistically stable changing frequency of flow directions, like the rotating frequency of the surface
vortex core for radial impellers, could not be found for PBT in the present work. The flow
directions can be completely opposite in a period regardless of the same rotation direction of
the stirrer, so that the surface vortices can appear before and behind a particular baffle from
time to time as shown in Figure 5.7.

5.1.4   Visualization of the surface vortex structure

5.1.4.1   Introduction and geometry

In order to study the structure and the
origin of the surface vortex, laser light
sheet visualization techniques were
applied in the present work. The visualizations were carried out in a
dished-bottomed   cylindrical   glass
vessel of diameter $T = 350$ mm. The
differences between the flow in a flat-
bottomed and a dished-bottomed tank
are negligible at the surface according
to Magni et al. (1988) [81]. Only the
radial impeller RT was involved since
the surface vortex in this case can be
captured easily during the investiga-

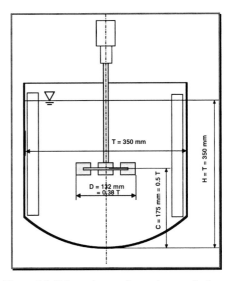

Figure 5.8: Schematic experimental set-up for laser
light sheet visualization for RT

tion. In order to investigate the influence of the baffles on surface vortex formation, the baffle number varied between 0, 2 and 4, corresponding to a baffle intensity number of 0, 0.15 and 0.26, respectively. The critical baffle intensity number for the RT was 0.22. Therefore, only the four-baffle configuration satisfied the fully baffled condition. The bottom clearance of the stirrer was chosen as $C/T = 0.5$, so that the vortex structure could be studied and compared both above and below the stirrer. The geometric parameters are shown in Figure 5.8. The visualization results were again recorded by a video camera and stored in a PC.

5.1.4.2   Results and discussions

First, the flow pattern in an unbaffled condition was visualized. As shown in Figure 5.9-a, a three-dimensional flow pattern is formed already at a very low rotational speed. In addition to the primary rotational flow pattern generated by the rotation of the stirrer, a secondary flow pattern comprising radial and axial components appears, which is of more importance for the mixing process. A typical radial pumping flow pattern can be recognized, the liquid is discharged in the form of radial jet stream toward the wall, and at the wall is split into two axial streams along the side wall. These two streams are drawn into the stirrer axially. Thus two large-scale circulation loops are formed. The flow shows no velocity gradient as viewed at the surface. The most interesting region is a columnar cylinder region corresponding to the forced vortex region mentioned in Section 5.1, and this region can be better marked in the visualizations as depicted in Figure 5.9-a where few radially and axially moving tracing particles are to be seen. The liquid in this region is clearly in a pure solid body rotation. This column also marks off the boundary of the two circulation loops. With increase in the impeller rotational speed, the surface changes its profile shape, until the depth of the central vortex reaches the stirrer region and brings air bubbles into the vessels.

After two and four baffles have been introduced into the vessel, the columnar solid body rotation region can no longer be seen. The inner boundary of the two large-scale circulation loops can reach the stirrer shaft directly. In comparison with the unbaffled vessel at the same N, the axial and radial components in the secondary flow pattern increase significantly in magnitude. Owing to the obstacle influence of the baffles, the circulation motion of the stirred fluid is disturbed and the formation of the central vortex is suppressed. The tangential component of the circulation flow motion is rebounded by the baffles, causing a high degree of turbulence. This allows a higher mixing efficiency in baffled tanks.

As the impeller rotational speed is increased to a certain level, the flow in the region surrounding the stirrer shaft is still in high and irregular rotational motion. Instead of the appearance of a central vortex, a small single vertical vortex appears in this region, and rotates around the stirrer shaft. Figure 5.9-b illustrates the structure of one surface vortex core at the instant when the vortex passed the plane emitted by the laser beam. Actually, this vortex core is the rotation centre of the whole flow, as already depicted in Figure 4.14 for rotating disk

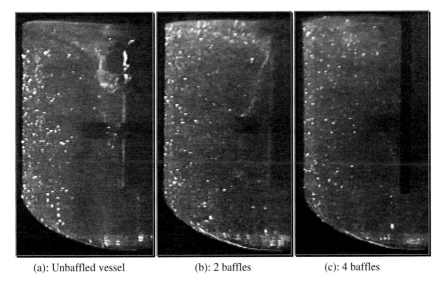

| (a): Unbaffled vessel | (b): 2 baffles | (c): 4 baffles |

Figure 5.9: Laser light visualization result at $N = 200$ rpm

flow in a square tank. The surface vortex is the appearing mode of the precessing central vortex at the surface. This new type of vortex does not change its surface shape as strongly as in a forced vortex motion. A high swirling intensity still remains at the core of the vortex. A concentrated Rankie vortex is formed in which the size of the forced vortex is very small. The core of this new type of vortex is put into a precessing motion around the geometry centre, namely the shaft of the stirrer. The off-set of the precessing vortex core is much larger in the stirrer region where a strongly radial jet exists. The comet tail behaviour of the small air bubbles mentioned in the last section is actually the motion following the precessing motion of the vortex core as depicted in Figure 5.9-b. This phenomenon is widely termed PVC (precessing vortex core), which was summarized as a common feature of highly swirling flow in Section 2.1.8. The stirrer and the radial jet separate this vortex core into the upper part (Figure 5.9-b) and the lower part (Figure 5.9-c) of the tank. These two parts seem to move with the same behaviour, but small differences can appear occasionally.

As the reasons, it was assumed in the present work that the vortex core centre is disturbed mainly by the obstacle influence of baffles and any other imperfection in the geometry such as eccentric shaft installation, etc. Among them, the baffle is of the greatest importance. In unbaffled tanks this phenomenon has not been found. As a consequence of these disturbances, the vortex core cannot maintain its original position and departs from the geometry axis in a precessing motion. If the disturbances supply enough energy, it can be relocated to the outer orbit in a spiral trace, and sometimes the precessing can disappear from its outer orbit to the shaft again along a spiral trace.

The baffle number has a significant influence on the intensity of the surface vortex and the precessing motion of the vortex core. As only two baffles were introduced in the tank, the surface vortex has a higher intensity and deeper penetration depth. Air bubbles are entrapped already at a relatively low impeller speed, indicating the non-fully baffled situation. With increasing baffle number, the vortex intensity, the penetration depth and the swirling intensity decrease. The precessing motion of the vortex core becomes more stable. Note that even with four baffles, namely under fully baffled conditions, this surface vortex is not totally eliminated as assumed by Liepe *et al.* [73].

It can be concluded that for radial impellers, the surface vortex is nothing else than an appearing form of the precessing central vortex core at the free surface. The vortex core is generated and disturbed by the presence of the baffles. The intensity of the vortex decreases with increasing baffle number.

5.2   Critical rotational speed for the onset of surface aeration and correlations

5.2.1   Definition of different critical rotational speed

In pre-observations, the rotational speed of the stirrer was first increased from zero to the maximum allowable level and then decreased back in the same manner to check the repeatability of the corresponding phenomenon observed. It was concluded that significant differences existed between the results of the two procedures. This can be attributed to the fact that during the decreasing procedure the already entrained bubbles still existed in the liquid. These bubbles influenced, on the one hand, the physical characteristics of the stirred liquid considerably. On the other hand, the judgement of the status of the aeration process was disturbed by these bubbles. Therefore, in the present work, all the critical rotational speeds were determined only in the upwind procedure. During the increasing process, each impeller speed was maintained for at least one minute to achieve a steady state.

According to the literature survey on surface aeration discussed in Section 1.2.2, four different criteria have been widely applied by various research groups. The first two were based on visual observation, namely the critical impeller speed N_{CSA1} for the first visible bubble entrained from the free surface, and the critical impeller speed N_{CSA2} for the steady state of surface aeration. The other two can be quantitatively determined by measuring the sudden drop in the power number Po and the sudden increase in the mass transfer rate of the gas phase. The last two criteria are normally determined in a fairly high gas hold-up state, with massive gas bubbles already aerated into the stirred liquid. Bubble entrainment starts at a low impeller speed prior to any drop in Po. In addition, because of the coarse resolution during the increasing impeller speed for determination of the power number, the sudden drop was not observed in the present work. Heywood *et al.* (1985) [59] also pointed out this type of difficulty. In the present work only the visual criteria were applied to determine the onset of surface aeration.

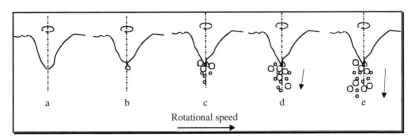

Figure 5.10: 5 sub-processes of the bubble entrainment in stirred vessels

For all types of stirrers, it was concluded that the surface bubble entrainment process in stirred tanks can be refined into five main steps according to our observations: the first visible surface vortex, the first single visible bubble (this state corresponds to N_{CSA1}), the cloudy bubble cluster at the surface vortex bottom (the state as more than five single bubbles accumulated at the vortex bottom in the present observation), the first visible bubble transferred into the stirrer region, and the steady state of surface aeration (corresponding to N_{CSA2}). These five developing sub-processes of surface aeration are represented schematically in Figure 5.10-a-e. The corresponding rotational speeds at these states were characterized as N_{vortex}, N_{bubble} (N_{CSA1}), N_{cloudy}, $N_{l.in}$ and N_{steady} (N_{CSA2}).

5.2.2   Comparison with the existing correlations

The critical rotational speed of the RT stirrer was determined, which was already applied for the elucidation of the mechanism of surface aeration. The observations were carried out both in water and in the silicone oil mixture. Since most correlations existing in the literature were deduced for water, only the results for water were presented and compared with the correlations. For each critical rotational speed, at least 10 observations were made to obtain an average value. The standard deviation was also estimated and normalized by the mean value of the corresponding critical rotational speeds. The results are listed in Table 5.3.

For all critical rotational speeds except N_{cloudy}, the deviations are in the range 16-20%, indicating a weak repeatability. Since the appearance of the first vortex, the presence of the first bubble and the first bubble in the stirrer region are purely random events and the corresponding critical impeller speeds are neither representative nor reproducible. Therefore, they are actually not appropriate for the onset of air entrainment. The criterion of the steady state of aeration is also very subjective and depends very strongly on the observer, hence representing a limitation for wider application. In contrast, N_{cloudy} had the smallest deviation (9.3%) in comparison with the others, indicating a more stable sub-process of air entrainment. However, a difficulty exists in the definition of the extent of the cloudy status. In the present work it was defined as the state in which at least five bubbles accumulated at the bottom of one surface vortex. In previous investigations, only Tanaka *et al.* (1986) [124], (1987) [123] applied this

	N_{vortex}	N_{CSA1}	N_{cloudy}	$N_{l.in}$	N_{CSA2}
Mean [rpm]	131.3	243.7	264.0	295.7	319.0
Standard deviation [%]	18.1	19.8	9.3	16.4	20.2

Table 5.3: Determination of different critical rotational speeds of onset of surface aeration

criterion. They did not explain the exact definition of the state. They also studied the relation between N_{cloudy} and N_{steady}, and it was reported that the latter is always about 19% higher than the former. In our case, this difference was about 20%, which agrees well with their results. Ditl et al. (1997) [34] did not present an exact criterion to define the critical impeller speed in their investigations, they only pointed out that this value was taken as air entrainment was visually observable. According to our experience, their value corresponds to the N_{cloudy}.

The critical rotational speeds determined by visual observations in the present work were also compared with the literature. Since in the present work only the criteria determined by visual observation were applied, the results were compared only with those of studies in which the similar visual criteria were applied. The results are presented in Table 5.4. The present work agrees well with the work of Heywood et al. [59] and Ditl et al. [34] in which a very similar geometry was applied for the RT. The deviations from the present work are only 4.8 % and 9.6 %, respectively. The correlation of Heywood et al. [59] is:

$$N_{CSA} = 1.04 T^{0.616} D^{-0.97} C^{-0.23} H^{0.59}, \qquad (5.1)$$

where T, D, C, H denote the diameter of the tank, the diameter of the stirrer, the bottom clear-

Authors	N_{CSA1} [1/min]	N_{cloudy} [1/min]	N_{CSA2} [1/min]	Deviation [%]
Present work	243.7	264.0	319.0	-
Dierendock	285.1		-	16.9
Heywood	234.3		-	4.8
Ditl		267.5		9.6
Tanaka		425.6	510.7	67.0
Greaves	-		380.8	19.4

Table 5.4: Comparison of N_{CSA} values in the present work and in the literature

ance of the stirrer and the filling height, respectively, and that of Ditl *et al.* [34] is

$$N_{CSA} = \sigma^{0.16}\nu^{0.17}T^{-1.09}\left(\frac{H-C}{H}\right)^{0.66}\left(\frac{T}{D}\right)^{1.58},\qquad(5.2)$$

where in addition to the geometric parameters also the physical parameters of the liquid, i.e. the surface tension and the kinematic viscosity, were taken into account, denoted by σ and ν, respectively.

Note that even though Tanaka *et al.* [123] applied exactly the same criteria as in the present work, considerable deviations in the two correlations exist. The reason might be attributed to the different definition of the state of cloudy clusters and the steady aeration, which are very subjective and depend strongly on the observer. For example, in the present work, the state at which at least five bubbles accumulate at the bottom of the surface vortex was taken as the critical rotational speed for cloudy enfolding, whereas in the work of Tanaka *et al.* [123] a more developed cloudy enfolding state might be achieved.

5.2.3   Theoretical analysis of the correlations

Although the main objective of the present work was not to set up a new correlation for the onset of surface aeration in stirred flows, the correlations which are based on more physical models were checked and the results from them were compared with the results of visual observations carried out in the present work.

Based on the visual observations, the surface aeration can be described as a consequence of the combination of different forces acting at the surface. First of all, owing to the gravity forces, the surface vortex has to change its shape to follow the pressure distribution; the Froude number Fr determines this process. The surface vortex is disturbed by the mean flow at the surface and the rebounding flow from the baffles, and hence becomes unstable. The Reynolds number Re plays an important role in describing the disturbance. Bubbles can be formed at the bottom of the unstable vortex owing to the enfolding effect, and stay at the vortex bottom until the downward velocity is high enough to overcome the surface tension and the rise velocity of the bubbles. Hereby the Weber number We is involved.

As the first step, we can assume that regular surface aeration occurs as the downstream velocity in the circulation flow can pull down bubbles from the surface. For low-viscosity fluids at higher Reynolds number, the axial downstream circulation velocity U_c near the surface in the circulation flow is proportional to the tip velocity of the stirrer blade V_{tip}. This gives

$$U_c \propto ND.\qquad(5.3)$$

The characteristic diameter of the bubbles at the beginning of the surface aeration can be estimated according to Wichterle *et al.* (1996) [138] as

$$d_B = \sqrt{\frac{\sigma}{\Delta\rho \cdot g}} \, , \tag{5.4}$$

where $\Delta\rho$ represents the density difference between the gas and liquid phase, and normally the density of the gas phase can be neglected, giving $\Delta\rho = \rho_L$. The rise velocity U_B of such bubbles can be approximated in low-viscosity liquids as

$$U_B = \left(\frac{\sigma \cdot g \cdot |\Delta\rho|}{\rho^2}\right)^{1/4} = \left(\frac{\sigma \cdot g}{\rho}\right)^{1/4}. \tag{5.5}$$

Assuming that the onset of surface aeration happens under the condition $U_c > U_B$, the critical dimensionless velocity for the onset of surface aeration, i.e. the ratio of U_c to U_B, can be expressed as

$$U^* \equiv \frac{ND}{U_B} = ND \cdot \left(\frac{\rho}{\sigma \cdot g}\right)^{1/4} = \left(We \cdot Fr\right)^{1/4}, \tag{5.6}$$

where the Weber number is $We = \dfrac{N^2 D^3 \rho}{\sigma}$ and the Froude number is $Fr = \dfrac{N^2 D}{g}$.

According to Equation (5.6), in a similar geometry configuration in stirred vessels, the minimal dimensionless velocity to generate surface aeration should be a constant.

The above assumption was verified by visual observations carried out in water and in the silicone oil mixture in the present work. The surface tension σ of the two media were measured as 0.073 and 0.023 *N/m*, respectively. The dimensionless critical velocity for N_{cloudy} was calculated as 5.18 for water and 5.15 for silicone oil mixture, and for N_{CSA2} 6.78 and 6.13. Good agreement between the two fluids was achieved. This agreement confirms the validity of the physical modelling by using dimensional analysis.

Taking into account the whole surface aeration process and using dimensional analysis in a similar way, the onset condition can be described by the function

$$f(Fr, Re, We) = 0 \tag{5.7}$$

Again, the dimensionless numbers are: $Fr = \dfrac{N^2 D}{g}$, $Re = \dfrac{ND^2}{v}$ and $We = \dfrac{N^2 D^3 \rho}{\sigma}$. All of these criteria contain the impeller rotational speed N and impeller diameter D. It is useful to rearrange these numbers to separate the physical parameters and the operation parameters:

$$Fr = f(We', Mo), \tag{5.8}$$

where We' is the modified Weber number in which the rotational speed N is filtered out. It can be expressed as

$$We' = \frac{We}{Fr} = \frac{D^2 g \rho}{\sigma}. \tag{5.9}$$

Mo represents the Morton number, containing only the physical parameters of the liquid

$$Mo = \frac{We^3 Fr^2}{Re^4} = \frac{g \rho^3 v^4}{\sigma^3}. \tag{5.10}$$

Ultimately, taking into account the geometry variations, the condition under which surface aeration begins can be expressed as

$$Fr = C_1 We'^{\alpha} Mo^{\beta} \left(\frac{H - C}{D} \right)^{\gamma} \left(\frac{T}{D} \right)^{\delta}, \tag{5.11}$$

where the exponents need to be determined for each geometry configuration separately. Ditl et al. [34] determined these exponents for different impeller types and configurations from visual observations. Similar dimensionless groups were also applied in the work of Zehner et al.[149]. However, in their correlations, only the power number drop criterion (N_{CSA3}) was used, preventing a quantitative comparison. For RT stirrers with different D/T ratios, the exponents determined by Ditl et al. [34] are listed in Table 5.5.

In order to verify the validity of the physical modelling of the surface aeration process, the critical rotational speeds for a group of RT stirrers were determined. The RT stirrers varied in the impeller diameter D, and hence the D/T ratios. Four RT stirrer elements were selected, and the detailed dimensions are listed in Table 5.6. The observations were carried out both in wa-

C_1	α	β	γ	δ
30.11	-0.59	0.088	1.31	0.97

Table 5.5: Values of parameters in correlation (5.11) by Ditl et al.

ter and in the silicone oil mixture. The N_{cloudy} criterion was used. The results obtained in the observations are compared with the predicted values from Equation (5.11). Figure 5.11 presents the comparative results.

D	D/T	D_{Disk}	D_{Hub}	D_{Shaft}	L_{Blade}	W_{Blade}	T_{Blade}
90	0.225	67.5	29.5	20	22.5	18	2
132	0.33	99	29.5	20	33	26.4	2
180	0.45	135	29.5	20	45	36	2
260	0.65	195	29.5	20	65	52	2

Table 5.6: Geometry (mm) of stirrer elements employed for the correlations (RT stirrers)

It can be concluded that very good agreement was achieved both for water and for the silicon oil mixture between the results of visual observations and the values predicted from physical modelling. Therefore, the correlation of dimensionless numbers can describe the surface aeration process more reasonably and accurately, and hence can be used to predict the critical impeller speeds for the onset of surface aeration during the design and optimisation of the stirring system.

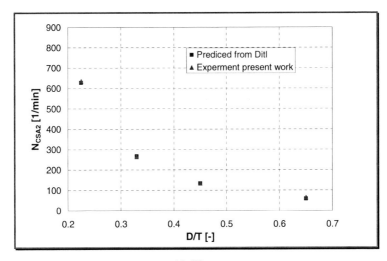

(a): Water

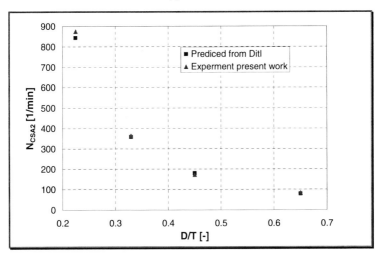

(b): Silicone oil mixture

Figure 5.11: Comparison of critical impeller speeds between predicted and experimental results

6 EVALUATION OF STUDIES ON SURFACE AERATION AND MACRO INSTABILITY

The steady flow pattern induced by standard stirring elements such as RT and PBT in baffled tanks was studied and characterized by numerous experimental and numerical studies. Various aspects of the flow were studied, e.g., mean flow velocities, turbulent velocity fluctuations, energy dissipation rate and time and space scales of turbulence. Intensive research [108] on flow fields in stirred vessels resulted in a wide knowledge of the stirring process, permitting effective design of the stirring process. The mean flow hydrodynamics in stirred tanks are quite well understood. However, the flow in stirred vessels is actually associated with unstable phenomena as demonstrated in the last two chapters, e.g., precessing vortex core motion symbolized by the surface vortex in stirred flows of RT, the non-rotational flow pattern varying in the flow direction for PBT, and swelling surges before baffles. The flow consists of a set of hierarchically organized unstable large-scale motions evidently differing from the pure turbulence eddies. These unstable motions have attracted increasing attention in recent years, and are widely termed macro instabilities (MI).

In this chapter, the steady flow pattern induced by the RT and PBT is first introduced as a background to the rotational flows in stirred vessels. Attention is also paid to the flow near the free surface. Subsequently, the long-time large-scale MI motions were characterized by velocity time series and spectral analysis. The magnitude of the MI contributions to the RMS value and the turbulence kinetic energy in stirred flows was estimated by a decomposition method. The influence of geometry on the dominant frequency was studied in the MI variation investigations. The swelling surges before the baffles intensify the unsteadiness of the flow at the surface, thus accelerating the air entrainment process at the bottom of surface vortices significantly. They are characterized by scaling of the wall jet flow along the baffles. Finally, two possible methods to suppress air entrainment are introduced and analysed.

6.1 Flow Pattern Investigation

6.1.1 Flow configuration

The LDA measurements of the stirred flow pattern were carried out in a cylindrical vessel of diameter $T = 400$ mm. The dimensions are listed in Table 3.1. The vessel had a free liquid surface. Again, two types of stirrer elements, the RT and the PBT, were considered, with a D/T ratio of 0.33. The geometry of the stirrers is presented in Figure 3.5 and the dimensions are listed in Table 5.1. The baffle number varies between 0 and 4. For the unbaffled and the two-baffle configurations, the flow field was measured only in one vertical plane, and this plane was located in the middle plane between two baffles for the latter case. For the four-baffle configuration, the flow field was measured in different vertical planes (different θ values before one baffle), as shown in Figure 3.4-b.

During the LDA measurements, the impeller rotational speed was chosen as the maximum allowable value at which no visible air entrainment could be observed. Table 6.1 lists the corresponding operating conditions for the RT and PBT in flow field investigations. In all cases listed in Table 6.1, the impeller Reynolds number did not exceed 10^4 owing to the relatively high viscosity of the working fluid used in the investigations. These values are much lower than the critical Reynolds number predicted by Liepe *et al.* [73] (see Table 2.2), indicating that a fully turbulent flow regime is not achieved either in the impeller region or in regions away from the impeller. Note that the flow at this Reynolds number is still turbulent. Schäfer *et al.* (1997) [110] and Dyster *et al.* (1993) [39] confirmed that both the flow velocity and its RMS value are directly proportional to the impeller tip velocity in the impeller region, indicating a weak Reynolds number dependence in this region. However, this proportionality is not valid in regions far from the impeller.

For the representation of the bulk flow pattern generated by the RT and PBT in different flow

Impeller type	Baffle No.	D [mm]	N [rpm]	Re [-]
RT	0	132	300	5424
RT	2	132	250	4520
RT	4	132	300	5424
PBT	4	132	405	7322

Table 6.1: Flow operating conditions of RT and PBT

configurations, the spatial coordinates r and z are presented in dimensionless form scaled by the tank diameter T. The radial, tangential and axial velocity components are normalized by the impeller tip velocity $V_{tip} = \pi DN$ as U/V_{tip}, V/V_{tip} and W/V_{tip}. Since the present work is also concerned with the rotational flow in stirred vessels, the tangential velocity component is also often expressed as the rotational speed as $n = W/2\pi r$ and normalized by the impeller speed N as n/N. In addition, the turbulence kinetic energy k used to estimated the spatial distribution of the turbulence level was calculated from the root mean squares of three velocity components u', v' and w' (RMS velocities) as

$$k = \frac{1}{2}(u'^2 + v'^2 + w'^2).$$ (6.1)

The turbulence kinetic energy is normalized by V_{tip}^2 and the corresponding values are denoted by k/V_{tip}^2. The normalization of all values simplifies the comparison between measuring results obtained with different vessel sizes and operating conditions.

6.1.2   Unbaffled and non-fully baffled stirred flows

Figures 6.1 and 6.2 show the flow field in the r-z plane of the RT in the unbaffled tank and in the tank with two baffles, respectively. In both figures, the vector represents the velocity projected into the r-z plane. The contour colours in the left figure represent the normalized turbulence kinetic energy, whereas in the right figure the contours represent the normalized rotational speed of the flow. The reference vector for the velocity representation and the scale of the contours are set with the same value for a convenient comparison between different flows.

In both cases, the secondary flow pattern consists of two circulation loops. However, the secondary flow pattern differs in intensity considerably in the two cases. For the unbaffled stirred flow, in the region near the shaft and the free surface no LDA measurement signal could be obtained owing to the presence of the central vortex. The radial jet flow is somehow inclined towards the lower circulation loops, with a maximum angle of about 34° from the horizontal line. This inclination was not observed in the experimental (Nagata *et al.* [89]) and numerical investigations (Ciofalo *et al.* [24]) in unbaffled stirred flows with and without a free surface. This phenomenon is attributed to the influence of the central vortex, while in the previous work the vortex was not so deep as to influence the trailing vortex in the discharge flow in the impeller region. The maximum radial velocity component in the radial jet flow is 0.22 V_{tip}. The location of the two circulation loop centres is at the point $r/T = 0.4$, $z/T = 0.4$ for the upper loop and $r/T = 0.35$, $z/T = 0.15$ for the lower loop. The upper loop centre is moved outwards owing to the presence of the central vortex. The upper circulation loop achieves up to $z/T = 0.65$, whereas the lower circulation loop covers the whole range below the stirrer. The radial and axial velocity components in regions far from the impeller are very small, whereas the tangential velocity component is much larger. The rotational speed of the flow shows a columnar distribution form. Almost no axial gradient of angular velocity exists in the tank. In the region $r/T < 0.1$, the fluid rotates almost with the same angular velocity as that of the stirrer ($n/N \approx 0.9$), whereras the fluid below the stirrer rotates slower than in the upper part (in the range $n/N = 0.68$-0.71). The angular velocity decreases quickly with increasing radial coordinate and approaches zero near the cylinder wall. The turbulence kinetic energy is relatively low even in the discharge flow. The maximum value in the vicinity of the stirrer is about $k/V_{tip}^2 = 0.05$. This value is smaller than 0.005 in the bulk of the tank.

For stirred flows with two baffles, similar circulation loops exist as in the unbaffled case. The secondary circulation loops increase in intensity and size significantly. The maximum radial velocity in the discharge flow increases to 0.46 V_{tip}, over double that in the unbaffled case.

The upper circulation loop achieves a much higher level ($z/T = 1$), very close to the surface. The circulation loop centre is located at point $r/T = 0.37$, $z/T = 0.53$ for the upper loop, and $r/T = 0.38$, $z/T = 0.24$ for the lower loop. Both loop centres are moved outwards in comparison with the unbaffled flow. The maximum turbulence kinetic energy in the impeller region is greater than $0.1\,V_{tip}^2$. Owing to the obstacle effect of the baffles, the angular velocity decreases strongly in regions far from the shaft. The $n/N > 0.2$ region shrinks to ca. $r/T = 0.26$, whereas this boundary is located at $r/T = 0.38$ for the unbaffled flow. The boundary between each level is not so well defined as the columnar distribution shape of the angular velocity in the unbaffled flow. The angular velocity can still achieve $0.8\,N$ in the region close to the shaft in the upper tank, indicating highly rotational flow in this region. The baffle effects is negligible in this region.

6.1.3   Fully-baffled stirred flow

6.1.3.1   RT

Because the baffles play an essential role during the formation of surface vortices, several measuring traverses were carried out at different position (r-θ planes) relative to the baffles shown in Figure 3.4 ($5°$, $15°$, $30°$, $45°$, $60°$, $75°$and $85°$ in front of the baffles; here only the planes $5°$, $45°$are shown).

Figures 6.3 and 6.4 present the typical mode of operation of the Rushton turbine (RT) in a fully baffled tank. The flow pattern in the middle plane retains a reasonable similarity to the flow field measured by Schäfer *et al.* [107] in a similar geometry but with a tank of different size. The liquid is drawn into the stirrer element axially, deflected in a radial direction and discharged again in the form of a radial jet stream. The radial jet splits at the vessel wall and is diverted upwards and downwards, leading to the formation of two large circulation loops, one above and one below the radial flow at the level of the stirrer blades. The centre of the upper circulation loop has a slight tendency to move downwards as the distance from the baffles increases except plane $5°$(from $z = 0.515T$ at $15°$ to $0.45\ T$ at $85°$). The upward spreading of the upper circulation loop decreases considerably from $z = 0.9\ T$ at $5°$ to $z = 0.68\ T$ at $75°$. The circulation penetrating height in the upper tank can also be determined by the condition that the axial velocity component is not less than $0.05\ V_{tip}$. In the plane shortly before the baffles (plane $5°$) the axial velocity achieves $0.05\ V_{tip}$ up to a height of $z = 0.93\ T$. This height decreases rapidly to $z = 0.75\ T$ at plane $45°$ and $z = 0.675\ T$ at plane $85°$. In all planes except plane $5°$, a secondary, and in some cases even a third, vortex appears in the uppermost area near the vessel wall. The centre of the secondary ring vortex is located at $z = 0.97\ T$, $r = 0.36\ T$ in plane $45°$.

Note that both axial and radial velocity components in all r-z planes are of very small magnitude at planes above $z/T = 0.8$. Little mixing takes place in this region. This also means that in all planes the active volume of the circulation flow can scarcely reach the fluid level. Similar results were also obtained by Bittorf and Kresta (2000) [8] by checking the similarity of the axial velocity profile before baffles of stirred flow induced by an axial impeller. They concluded that only a height of $z/T = 2/3$ can be achieved for axial impellers. However, this result contradicts the intense surges observed visually before baffles, and will be discussed further in Section 6.3.

The spatial distribution of the angular velocity remains in a columnar form. However, the region in which the angular velocity n/N is greater than 0.2 shrinks further, down to $r/T = 0.16$, indicating the intensified obstacle effect with increasing baffle number. The fluid in the vicinity of the impeller rotates with 0.6-0.8 N, whereas the maximum angular velocities reach about 0.25 N in the bulk flow and are limited in the range $r/T < 0.08$, where the fluid is more influenced by the rotation of the shaft. A swirl flow pattern exists in this range owing to the downwards and inwards fluid motion. In comparison with the flow near the wall, the obstacle effect of baffles in this range is negligible, since no dependence of flow on baffle position can be found.

Since the surface vortex is a type of vertical vortex, the flow field in a horizontal r-θ plane above the impeller is of more interest in the present work. LDA measurements were carried out with a finer grid in both circumferential and radial directions. In Figure 6.5 the flow pattern in two different z/T positions is shown: one is at $z/T = 0.75$ which is still within the upper circulation loop, and the other $z/T = 0.975$, i.e., just below the free surface. Compared with the flow in the discharge flow where the maximum radial component is over 0.5 V_{tip}, the mean velocity in plane $z/T = 0.75$ has a significantly smaller magnitude (with maximum velocity component about 0.1 V_{tip}). The tangential velocity component is much larger than the other two components: it is 5-6 times larger than the radial component (ca. 0.04 V_{tip}) and 10-15 times larger than the axial component (ca. 0.01 V_{tip}). The flow can be classified into three different regions. First, in the central region with a size close to the impeller, the fluid is in a purely tangential motion. The radial velocity component is nearly zero. Second, there is a small recirculation flow region in the vicinity behind the baffles with a size comparable to the baffle width. This zone extends to about 10° behind the baffles where the flow reattaches to the wall. The third region connects the two regions mentioned above. The flow convects radially inwards to the shaft owing to the redirection effect of the baffles, and this radial inward flow joins the centre region in a spiral manner. In plane $z/T = 0.975$, a very different flow pattern can be observed compared with that in plane $z/T = 0.75$. The tangential velocity component remains on the same level, whereas the radial velocity component is very small except in the vicinity of baffles. Purely rotational flow region expands in the whole plane, indicating invalidation of the influence of baffles at a height above the circulation loop range.

The influence of the baffles can be better understood by a comparison of the tangential velocity components between different baffle numbers depicted in Figure 6.6. For unbaffled stirred
flows, a typical Rankie vortex structure can be found from the tangential velocity component.
With the range $r/T < 0.1$ ($r/R = 0.6$), a solid body rotation motion ($V = $ constant $\cdot r$) is exhib-
ited with approximately the same angular velocity as the stirrer. Out of this region, a free vortex is formed ($V = $ constant$/r$), expanding to the vessel wall where the angular velocity is
close to zero. As two baffles are introduced, the tangential velocity decreases significantly.
The velocity profile shape, however, is similar to that of unbaffled flows, corresponding to the
non-fully baffled configuration with two baffles for RT. With four baffles, the tangential velocity profile keeps an approximately flat shape. The flow is then fully-baffled. The magnitude of the tangential velocity component is much lower than that of two-baffled and unbaffled configurations.

No vertical surface vortex structure can be found directly from the discussed steady flow pattern either in vertical r-z planes or in horizontal r-θ planes. The surface vortex is not developed from the mean flow in stirred vessels. In addition, as shown in Figure 6.6, the rotating
frequency obtained with visual observations does not coincide with the flow rotation. In the
range $r/T < 0.15$ where the surface vortex appears more often, the rotational frequency of this
vortex is much lower than that of the flow. This confirms the observation that the surface vortex does not move with the rotational flow and has its own regular rotation motion.

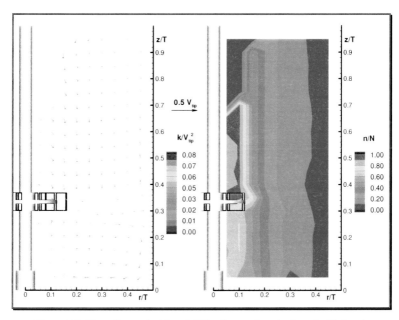

Figure 6.1: Mean flow velocities of RT in the *r-z* plane in the unbaffled tank

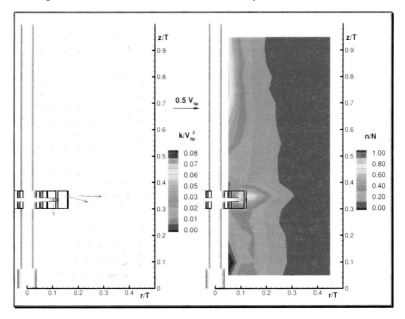

Figure 6.2: Mean flow velocities of RT in the *r-z* plane in the tank with two baffles

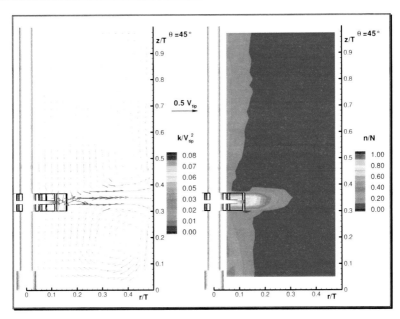

Figure 6.3: Mean flow velocities of RT in the *r-z* plane $\theta = 45°$ in the fully-baffled tank

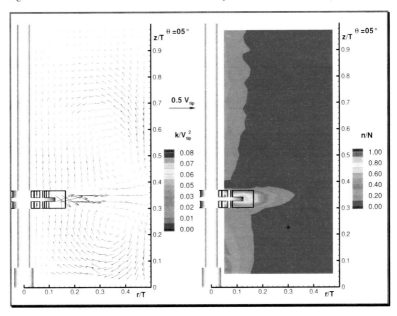

Figure 6.4: Mean flow velocities of RT in the *r-z* plane $\theta = 5°$ in the fully-baffled tank

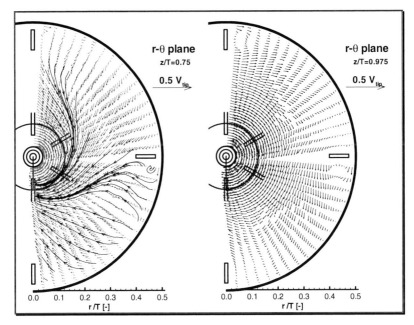

Figure 6.5: Mean flow velocities of RT in the r-θ planes in the fully-baffled tank

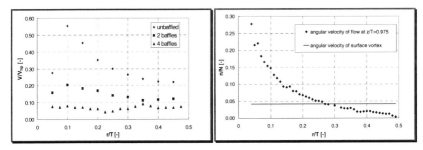

Figure 6.6: Comparison of tangential velocity component comparison with different baffle numbers

Figure 6.7: Comparison of angular velocities comparison between stirred flow at the surface and the rotational frequency of surface vortex

6.1.3.2   PBT

The mean velocity vector field and the angular velocity contours in the middle plane between two adjacent baffles ($\theta = 45°$) and also in the plane just before the baffle ($\theta = 5°$) are presented in Figures 6.8 and 6.9, respectively. Differing from the discharge jet stream generated by radial impellers, it originates from the lower edge of the impeller blade axially downwards with a peak value of about 0.5 V_{tip}, close to the value reported by Kresta and Wood (1993) [67] and Schäfer *et al.* (1998) [110]. The axial discharge jet spreads radially as it progresses towards the vessel bottom by entraining fluid from the surrounding jet. After the impingement of the jet on the bottom, the flow moves radially towards the side wall of the vessel, at which it turns upwards in a well-defined wall jet. As a direct consequence of the impingement at the bottom, the kinetic energy decreases considerably, from $0.03 V_{tip}^2$ shortly before reaching the bottom to ca. $0.01 V_{tip}^2$ in the radially directed jet flow. Unlike the RT, only one circulation loop is formed in flow induced by the PBT. The extending height of the circulation loop shows a strong dependence on the baffle locations, although the circulation loop centre stays nearly at the same position ($r/T = 0.35$, $z/T = 0.22$). In plane $\theta = 45°$, the circulation extends up to a height of $z/T = 0.6$, whereas this value increases to $z/T = 0.8$ in plane $\theta = 5°$. In the uppermost part above the extending height of the circulation loop, the flow is very weak and fairly irregular.

The spatial distribution of the angular velocity shows a different conical shape from the columnar shape of the RT. Accompanying the discharge jet out of the impeller, the angular velocity reaches a maximum value of ca. $0.45 N$ in the impeller region. This value is much smaller than that of the RT. Owing to the wall influence of the vessel bottom, the angular velocity decreases very quickly to $0.1 N$ just before reaching the vessel bottom. In most bulk flow regions, the angular velocity is smaller than $0.05 N$ except in the vicinity of the shaft. The rotational flow induced by the PBT is not uniform in different vertical levels and therefore more complicated than the rotational flow generated by the RT.

The flow field in the horizontal plane at two different vertical levels is shown in Figure 6.10. Within the circulation loop for example at level $z/T = 0.475$, the flow pattern is similar to that of RT. The flow pattern is distributed in three different regions: the redirected radial inward flow by baffles, the central pure rotational flow region with very small radial velocity component, and the flow region before the baffle. However, the spatial scale of the purely rotational flow shrinks to a radius $r/T = 0.1$, smaller than that of the RT. In the planes above the circulation expanding height, the flow is only active in the region behind baffles, for example at $z/T = 0.75$. In the region close to the wall, the magnitude of all three velocity component is very small. At even higher levels, the flow is totally irregular and non-active.

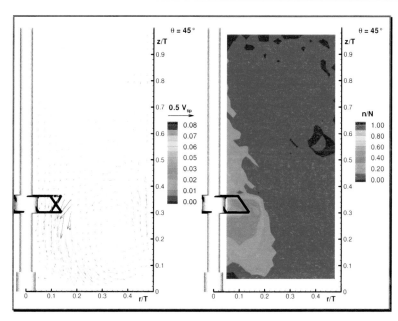

Figure 6.8: Mean flow velocities of PBT in the r-z plane $\theta = 45°$ in the fully-baffled tank

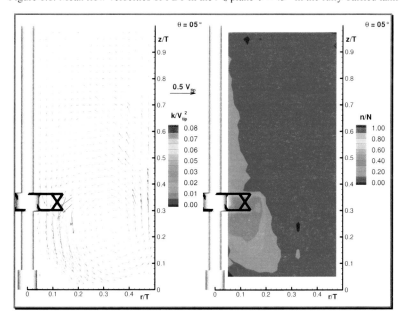

Figure 6.9: Mean flow velocities of PBT in the r-z plane $\theta = 5°$ in the fully-baffled tank

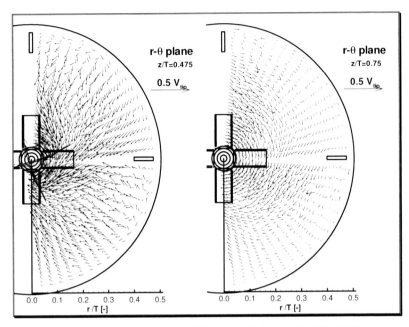

Figure 6.10: Mean flow velocities of PBT in the r-θ planes in the fully-baffled tank

6.2   Investigations of Macro Instabilities

As discussed in the last section, neither the dominant surface vortex for RT nor the dominant flow direction at the surface for PBT can be captured in the steady flow field investigation. The surface vortices are instantaneous phenomena, whose information is overwhelmed during the averaging process of the velocities. It was shown by Alekseenko *et al.* (1999) [2] that for well defined helical vortices but with a pronounced precessing vortex core, the average velocity profile can still obey the original swirling flow structures in which the precessing vortex core is embedded. Therefore, steady flow field investigations cannot characterize the precessing vortex core motion appropriately. Moreover, some instantaneous phenomena observed in the last chapter, such as wall jet flows before baffles whose intensity varies irregularly with time and also the asymmetric flow pattern at the surface, contradict the steady flow pattern significantly. These phenomena may be associated with the macro instability which has been intensively studied in the last decade. In this section, the characteristics of the MI are considered quantitatively based on the present LDA measurements.

6.2.1 Times series analysis of velocities

It was widely acknowledged in previous LDA measurement investigations in stirred vessels that the flow can be classified into two regions: the impeller discharge flow and the bulk flow [67,68, 107-112]. The velocity recorded in the discharge flow comprises the periodic oscillation induced by blade passages. The flow is highly periodic and can be better analysed by angle-resolved measurements. Trailing vortices are generated directly behind the blades and decay very quickly with the distance from the stirrer blades. The periodic oscillation no longer exists in the bulk flow any more [108]. The flow in the bulk region is considered to be steady [108], and purely embedded by random small-scale turbulent fluctuations for turbulent stirred flows. However, this assumption cannot be proved if the velocity time series at any point in the bulk flow is analysed in detail. Large-scale low-frequency variations are evidently superimposed on the pure turbulence fluctuations.

RT

Figure 6.11 shows the velocity time series at different points in the stirred flow induced by the RT. Two points were selected for the presentation. One is in the uppermost part of the tank close to the shaft where the surface vortices often appear, and the tangential velocity component is presented. The other point is located in the discharge flow, presented with the radial velocity component. Both in the bulk flow and in the discharge flow, quite well organized long-time oscillations appear in the velocity time series, and the time scale of these oscillation obviously exceeds the time scale range of pure turbulence fluctuations. These low oscillations are very similar to the PVC low-frequency variations depicted in Figure 2.15. Similar observations were also described by Kresta and Wood (1993) [68], and Myers *et al.* (1997) [87].

Compared with the mean value of 0.07 V_{tip}, the momentary velocity varies between -0.1 and 0.2 V_{tip}, indicating that the flow near the surface even changes its rotation direction regularly (Figure 6.11-a). At some instants the velocity can reach 0.46 V_{tip}, implying that the surface vortex is sweeping or passing through the measuring point exactly at these instants. The pe-

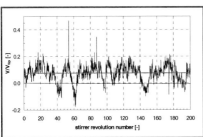
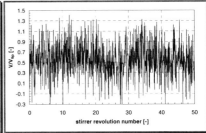

(a): V at $r/T = 0.07$, $z/T = 0.95$ (b): U at $r/T = 0.175$, $z/T = 0.35$

Figure 6.11: Velocity time series at a point in the bulk flow and in the discharge flow of RT

riod of the fairly well organized oscillations lies in the range of 20-30 stirrer revolutions (corresponding to 4-5 s), excluding the origin from the blade passage. In the discharge flow, the velocities are much higher than in the bulk flow. The velocities often exceed the tip velocity of the blades, induced by the strong trailing vortex motion in this region. The high oscillation with a period of approximately one stirrer revolution corresponds to a single blade passage. However, as shown in Figure 6.11-b, this passage periodic fluctuation (in the range of 1/30 s) is also superimposed upon the same low-frequency long-time oscillations. This type of low-frequency oscillation intensifies the variation of the velocities, leading to the appearance of velocities higher than the tip velocity of the impeller blade. These low-frequency variations occur in all three velocity components. Among them, the axial velocity component has the weakest variation amplitude.

PBT

For PBT, the long-time variations are much more complicated. Figure 6.12-a presents the velocity time series of the axial component at a point near the surface. The velocity fluctuates around zero, indicating the weak mixing efficiency in the uppermost part of the tank. The amplitude of the variations varies between -0.08 and 0.08 V_{tip}. Different periods (from ten rev to several hundred rev) appear in the time series, and they dominate with different levels in different time intervals. This low-frequency variation can be better revealed by the other two horizontal components. Figure 6.12-b presents the velocity time series of the tangential component at the same point. The velocity varies from -0.18 up to 0.2 V_{tip} with different variation periods. For example, a long period (ca. 300 rev) can be observed in Figure 6.12-b with a positive tangential velocity component. This indicates that the flow can keep in a rotating direction for 40-60 s. Periods of 10-100 rev are also obvious in the time series. These irregular variations agree well with the visual observation results for PBT that the dominant flow direction has a random period. Similar low-frequency variations appear also in the discharge flow induced by PBT.

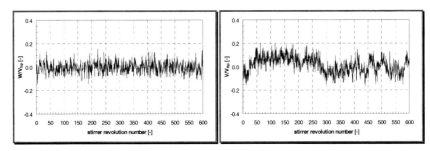

(a): W at $r/T = 0.125$, $z/T = 0.95$ (b): V at $r/T = 0.125$, $z/T = 0.95$

Figure 6.12: Velocity time series at point in the bulk flow and in the discharge flow of PBT

6.2.2   Spectrum analysis of macro instabilities

The long-time variations can be better analysed and characterized in the frequency domain. It is well known that the flow in stirred vessels consists of a set of hierarchically organized unstable formations such as eddies and vortices, e.g. trailing vortices induced behind the impeller blade, having length scales of the order of the blade height. The turbulent motion is usually described as a composition of eddies of different time and length scales [56]. These scales are often characterized by spectrum analysis of velocity and turbulent kinetic energy. In a one-dimensional velocity spectrum analysis for stirred flows, it was often observed that dominant frequencies appears in the lower range [102, 137]. No detailed explanation was given for these low frequencies, and this type of low frequency has been associated with the MI motion in the last decade.

6.2.2.1   Lomb periodogram algorithm

One of the most common standard spectral methods is the FFT (fast Fourier transform). However, the LDA measures the velocity of the seeding particles passing through the measuring control volume. The process of particle arrival is random. Therefore, the signal processing time interval is not constant and follows a Poisson distribution. The non-uniform distribution of the time between the samples precludes the use of the standard spectral estimators which are based on equidistant time intervals between samples. As an alternative, the irregular data can be resampled with an evenly sampled sequence using one of various interpolation techniques. Subsequently the familiar and accurate equidistant signal processing algorithms such as FFT can be used for the evaluation. This modified spectra method has been very widely used in turbulent spectral analysis in stirred flows, such as by Rao and Brodkey (1972) [102]. It was shown by Broersen *et al.* (2000) [13] that the resulting spectrum obtained by resampling suffers from a low-pass filtering effect, which cuts off the upper frequency band and causes a distortion of the estimated spectrum.

The Lomb-Scargle approach (hereafter abbreviated as the Lomb method) uses the precise time information in computing spectra and is a useful method for the detection of harmonic peaks. Therefore, it is specially appropriate for unevenly sampled signals and for the resonant frequencies in the lower range of the frequency domain. The method was originally proposed by Lomb (1976) [76] and has been widely used for time series analysis in astronomy to detect periodicity in data. The method is powerful in finding narrow peaks with a low noise level, and is especially developed for the detection of peaks. More detailed comparisons between the Lomb method and other approaches to the spectral analysis for unevenly sampled data can be found in Broersen *et al.* [13]. Details of the principle and the program of this algorithm are given in Appendix C.

RT

The power spectrum of the time series at two points, one in the bulk flow near the surface and the other in the impeller discharge stream, is depicted in Figure 6.13-a and b, respectively. It can be seen clearly that there is a peak frequency f_{MI} in the range of 0.2 Hz. Additionally, in the lower range of the frequency domain (0.1-1 Hz), there is a broad low-frequency band whose spectral power is much higher than that of the higher frequency band, indicating the existence of a series of low-frequency large-scale macro instabilities. No harmonic or other relationships between these low frequencies can be found. As shown in Figure 6.13-b, in the impeller discharge flow, the most pronounced frequency is much higher (ca. 29.9 Hz) with a high power density, corresponding exactly to the blade passing frequency (abbreviated as

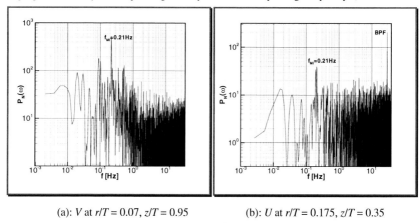

(a): V at $r/T = 0.07$, $z/T = 0.95$ (b): U at $r/T = 0.175$, $z/T = 0.35$

Figure 6.13: Frequency spectrum at points in the bulk flow and in the discharge flow for RT

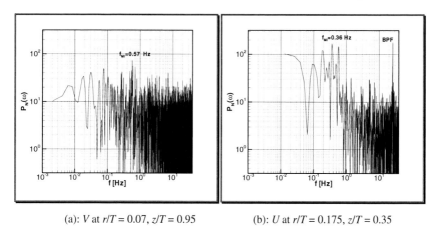

(a): V at $r/T = 0.07$, $z/T = 0.95$ (b): U at $r/T = 0.175$, $z/T = 0.35$

Figure 6.14: Frequency spectrum at points in the bulk flow and in the discharge flow for PBT

BPF in the figure): blade number $\times N = 6 \times 5 = 30$ Hz. This agreement verifies the prevalence of the trailing vortex motion in the discharge flow and furthermore confirms the accuracy of the spectrum evaluation of the Lomb algorithm. Note that even in this region, the lower MI frequency of 0.2 Hz is still pronounced, confirming the velocity time series analysis in the impeller discharge region.

PBT

No uniform dominant frequency can be found in the velocity spectra of PBT. A wider low-frequency band with approximately equivalent power levels appear in the spectrum shown in Figure 6.14. The blade passage frequency is likewise evident in the discharge flow (Figure 6.14-b). The scattering feature of PBT can be attributed, to some extent, to the measuring time sensitivity of the power spectral analysis for PBT, which will be discussed further in Section 6.2.2.5.

6.2.2.2   MI dominant frequency spatial distribution

RT

Based on the visual observations discussed in Chapter 5, the whole fluid content varies in the tank with a common rhythm with the precessing surface vortex. It can be imagined that the dominant frequency exists in the whole tank. In order to investigate the spatial distribution in the whole tank, the velocity spectra of all three velocity components in the whole tank were evaluated from the LDA measurements for the RT. For the presentation, the radial velocity component was selected. Figure 6.15-a illustrates the spatial distribution of the dominant frequency in a vertical r-z plane (plane 45°). Clearly uniform MI dominant frequency exists in the whole tank, except the impeller vicinity, where the MI dominant frequency is in the range of the blade passage frequency. This region spreads to about $r/T = 0.25$ in the radial direction and in the axial direction in the range $z/T = 0.275$-0.40, corresponding to the extent of the trailing vortex system. The flow in this region is highly periodic owing to the effect of the blade passage. In most of the rest of the tank, the MI dominant frequency is in the range of 0.2 Hz. Relatively greater deviations appear in the centre zone of both upper and lower circulation loops, indicating the additional unsteady circulation loop centre locations.

The MI dominant frequency also has uniform distribution in any horizontal plane. Figure 6.15-b presents the distribution in a plane near the surface ($z/T = 0.975$). In the circular range which does not expand to the baffle section, a fairly uniform distribution of the MI dominant frequency in the region of 0.2 Hz can be clearly observed. Within the baffle section, especially in the region close to the wall, owing to the impingement of the flow with the baffles and walls, the distribution is far from the uniform. Furthermore, this deviation is greater in the region behind the baffles than in the region before the baffles. This can be attributed to the fact that the MI is broken down by small-sized turbulent eddies due to the wall boundary or

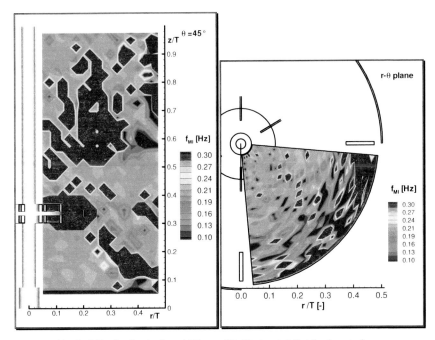

(a): Vertical distribution in the middle   (b): Horizontal distribution at plane near
plane between two baffles ($\theta = 45°$)   the free surface ($z/T = 0.975$)

Figure 6.15: MI dominant frequency distribution of radial component in the whole tank

the tail vortices behind baffles. This phenomenon was also mentioned by Hasal *et al.* (2000)
[58] for the stirred flow induced by a 45° six-blade PBT.

Despite of the deviations mentioned above, the spatial distribution of the MI dominant frequency can be seen as uniform in the whole stirred vessel. The whole flow field in stirred vessels varies its flow pattern with a common rhythm. This confirms the conclusion that the MI
motion is generated by the precessing vortex core motion.

PBT

There is no uniform dominant frequency in the stirred flow induced by the PBT. Figure 6.16
describes the dominant frequency and also the corresponding spectral power of the tangential
velocity component at profile $z/T = 0.95$. The dimensionless dominant frequency scatters in
the range 0.002-0.005. Even at the same point but measured at different time instants, this
dominant frequency also scatters. However, the spectral power (at most points above 1500)
remains on a much higher level than that of RT (with power in the range 500-1000) with the
same measuring time, implying a higher intensity of the low variation motions for PBT. The

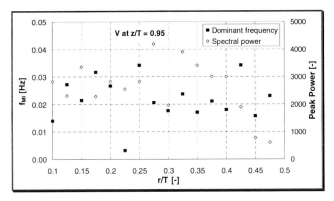

Figure 6.16: Spatial distribution of dominant frequency in the bulk flow for PBT

varying dominant frequency indicates a more complicated superimposition of different rotations with various dominant levels at different spatial points and time instants, thus resulting in different precessing speeds of the vortex core. The different angular velocity distribution depicted in Figure 6.8 confirms this assumption. The flow varies its flow field pattern with varying frequency. Similar results were also obtained in visual observations.

6.2.2.3   Linearity MI dominant frequency

The Strouhal number is often used to characterize unsteady flows. For PVC motion, the Strouhal number is defined by the precession frequency and the characteristic length and velocity as in Equation (2.23). For PVC in a stirred tank, the characteristic length can be taken as the impeller diameter D and the velocity as the impeller tip velocity V_{tip}. The Strouhal number can be rewritten for stirred flows as

$$Sh = \frac{f_{MI} \cdot D}{V_{tip}} = \frac{f_{MI} \cdot D}{\pi DN} = \frac{1}{\pi} \cdot \frac{f_{MI}}{N}. \qquad (6.2)$$

For investigations on macro instabilities in a stirred tank, the dominant macro instability frequency f_{MI} is more often normalized by the impeller rotational speed N as a dimensionless frequency f_{MI}^{*}:

$$f_{MI}^{*} = \frac{f_{MI}}{N}. \qquad (6.3)$$

This dimensionless MI dominant frequency f_{MI}^{*} can be considered as a modified Strouhal number for MI investigations in stirred flows.

As mentioned by various authors in MI investigations in stirred flows [14-16], the MI dominant frequency (for the case of an existing uniform dominant frequency) can be linearly related to the impeller rotational speed; as the impeller speed is increased to a certain level, a
constant dimensionless dominant frequency is acquired. Figure 6.17-a confirms this conclusion. A profile of the radial velocity component under the RT ($z/T = 0.175$) was selected for
analysis, where the dominant frequency has the highest spectral power. Linearity is fairly evident as the impeller speed was increased to above 150 rpm, verified by the high correlation
coefficient ($R^2 = 0.99$). Thus, the dimensionless MI dominant frequency f_{MI}^* equals 0.042 for
RT with the so-called standard configurations. Note that this constant dimensionless dominant
MI frequency coincides with the dimensionless rotating frequency of the surface vortex in-
cluded in the visual observations described in Section 5.1.3. This agreement connects the precessing vortex core motion and the MI phenomena in stirred flows. Therefore, it can be concluded that the MI motion in stirred vessels is nothing else than the consequence of the precessing vortex core motion superimposed on the rotational stirred flows. The PVC is the origin of the MI phenomena in stirred flows.

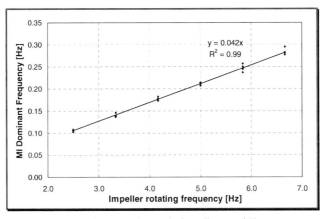

(a): Linear relation to the impeller speed N

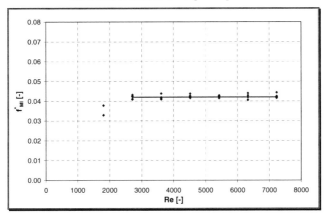

(b): f_{MI}^* against Re in a moderate range

Figure 6.17: Relationship between dominant frequency and operating conditions

Since the rotational speed of the motor driving the stirrer was very unstable in the low range, it was impossible to determine the onset of the PVC motion and the validity of the linearity of f_{MI} in stirred vessels in this range. As shown in Figure 6.17-b, the modified Strouhal number f_{MI}^* shows a very weak dependence on the Reynolds number at $Re > 2700$. As the For PBT this critical Reynolds number was estimated by Bruha *et al.* [15] as 4200, much higher than that of RT. Since the dominant frequency for PBT scatters strongly in a low-frequency band, a direct linear relationship is difficult to define. Nevertheless, in order to permit comparisons of dominant frequencies between the very different geometry configurations, the dimensionless dominant frequency is also used for PBT.

For stirred flows in similar geometries, the modified Strouhal number f_{MI}^* shows very weak dependence on the tank size and the working fluid in a moderate Reynolds number range. The f_{MI}^* was determined at the same locations ($z/T = 0.175$) in three similar geometry configurations which varied in the tank size and in the working fluid. For comparison, LDA measurements were carried out in tanks of diameter both 400 and 152 mm introduced in Table 3.1. In addition, a new working fluid, dimethyl-sulfoxide (DMSO), was used, which has the same refractive index but much lower viscosity than the silicone oil mixture applied in the present work. The results are summarized in Table 6.2.

T [m]	D [m]	D/T	Stirred medium			N [1/min]	Re	f_{MI} [Hz]	f_{MI}^* [-]
			Type	μ [mPa s]	ρ [kg/m^3]				
0.4	0.132	0.33	Silicone oil	16.4	1024	300	5424	0.21	0.042
0.152	0.05	0.33	Silicone oil	16.4	1024	1500	5188	1.27	0.038
0.152	0.05	0.33	DMSO	2.14	1100	300	6425	0.20	0.041

Table 6.2: Comparison of f_{MI}^* between different tank sizes and applied media

It can be seen that f_{MI}^* varied slightly in the range 0.038-0.042. Therefore, the Strouhal number can be considered in a moderate Reynolds number range as a constant within the precision of both experiments and evaluations. However, if the Strouhal number f_{MI}^* is observed in a wide range of Reynolds number, weakly dependence can be found as shown in Figure 6.18. Thanks to the relatively low viscosity of the DMSO, the modified Strouhal number f_{MI}^* was also measured in a higher Reynolds number region up to $Re = 29,000$. The Strouhal number decreased slightly with increasing in Reynolds number, indicating that the precessing rotational frequency cannot synchroly follow the increase of the rotation of the flow. This weak dependence explains, to some extent, the differences between different dominant frequencies reported by various workers. For example, a similar f_{MI}^* value (0.037) can also be recalculated from the work of Haam and Brodkey [56] at $Re = 19,000$ for a similar geometry but with water in a cylinder with elliptical dished bottom. Wernersson and Trägärdh (1998) [137] likewise obtained a dominant lower frequency in the turbulence kinetic energy spectrum analysis in a similar stirred vessel equipped with two RTs. The impeller spacing was 0.7 T. By careful rearrangement of their experimental results, an f_{MI}^* of 0.047 can be estimated from their values.

A lid located on the liquid surface has no influence on the dominant frequency. For the first two cases in Table 6.2, a surface lid was installed in the tank. The dominant frequency remained at the same value as that without a lid, indicating that a lid on the top of the tank cannot eliminate the precessing vortex core motion in stirred tanks.

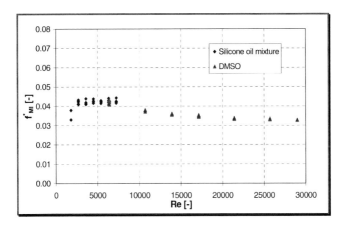

Figure 6.18: f_{MI}^* against Re in a wide range

6.2.2.4   Effect of sampling rate on MI

Since the measuring rate of LDA varied from 100 up to 400 Hz depending on the location of the LDA measuring volume in the vessel, the influence of the measuring rate on the determination of the MI dominant frequencies was investigated. Several measuring points in the stirred flow induced by the RT were selected. The measuring rate was adjusted individually from 50 to 500 Hz. The measuring time, however, was kept constant for each measuring rate.

The spectral power of the dominant frequency increases with increasing measuring rate, resulting from the increase in the sampling numbers in the Lomb algorithm. For better comparison, the spectral power of the corresponding frequencies was scaled by the maximum of the spectral power and expressed as a percentage, as shown in Figure 6.19. Only small deviations

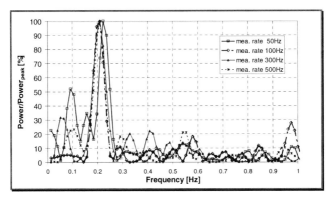

Figure 6.19: Spectral analysis comparison between different measuring rates

appear in the spectrum. Comparative results between different measuring rates at point $r/T = 0.2$, $z/T = 0.95$ are, for example, shown in the figure. Except for that the dominant frequency ranges are independent of the measuring rate for the lowest measuring rate of 50 Hz. Therefore, the effect of the measuring rate on the determination of the dominant frequency can be neglected. During the LDA measurement for MI analysis, the measuring rate was kept carefully in the range 100-300 Hz.

6.2.2.5   Effect of the sampling time on MI

Keeping the measuring rate constant, spectral analysis applying the Lomb method was carried out for different measuring times. For the present work, this was realized by taking different data numbers in the time series. The aim was to determine an appropriate uniform measuring time for all points during the spectral analysis procedure.

For the RT, no evident difference can be observed except for an increase in the spectral power of the dominant frequency with increasing measuring time. Figure 6.20 presents a comparison between the different measuring times for the same point in the spectral analysis. The peak arises clearly around 0.21 Hz for all cases. The noise for all evaluations remains at the same level (the spectral power is below 100). No additional peak exists in the spectrum. Therefore, for the RT the measuring time has no significant influence on the determination of the dominant MI frequency peaks.

For the PBT, the influence of the measuring time is much more complicated than that of the RT. The peak in the frequency domain appears in different ranges depending strongly on the measuring time. Figure 6.21 illustrates this dependence. As shown in Figure 6.21-a, for a measuring time of 2 min, the dominant frequency lies in the range 0.56-0.6 Hz (corresponding to an f_{MI}^* of about 0.01) with the second and third peaks appearing in nearly harmonic form. This peak frequency is very close to the result of Roussinova and Kresta (2000) [103]. They applied a similar geometry for PBT and obtained a dominant frequency of 0.62 Hz at 400 rpm with a measuring time of several seconds. Montes *et al.* (1997) [86] obtained a much higher f_{MI}^* of 0.06, but for a slightly higher impeller clearance of $C/T = 0.35$. When the measuring time was increased to 5 min, as depicted in Figure 6.21-b, the first peak moved to about 0.086 Hz, accompanied, however, by a second peak at 0.6 Hz. At the same time, a much lower peak around 0.028 Hz, emerged. With longer measuring times up to 10-20 min (Figure 6.21-c and d), the dominant frequency peak stabilized in the lower band range, namely 0.028 Hz ($f_{MI}^* = 0.004$). Myers *et al.* (1998) [87] investigated the dominant frequency in a similar geometry by applying PIV investigations for at least 20 min at one measuring point, and they also obtained a value nearly in the same range ($f_{MI}^* = 0.007\text{-}0.011$).

It can be concluded that for PBT, since the dominant frequencies occur in a very low band range, the measuring time plays the essential role in the Lomb method. This dependence can be used to explain the scattered dominant frequencies in the spectral analysis for PBTs. As shown already in Figure 6.16, even when the measuring time has been controlled carefully at a fixed value, the dominant frequency scatters in a close low range. This scattering indicates that the flow field of the PBT is more unstable and has inherently different dominant frequencies, which are mixed together with varying contributions to the macro instability motions.

Based on the above investigation, the measuring time both for the RT and the PBT were maintained at more than 15 min during the measurements for spectral analysis.

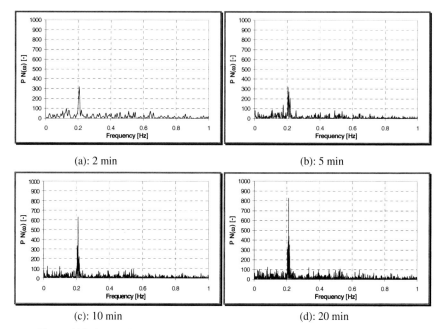

(a): 2 min   (b): 5 min

(c): 10 min   (d): 20 min

Figure 6.20: Comparisons of spectral analysis with different measuring times for RT

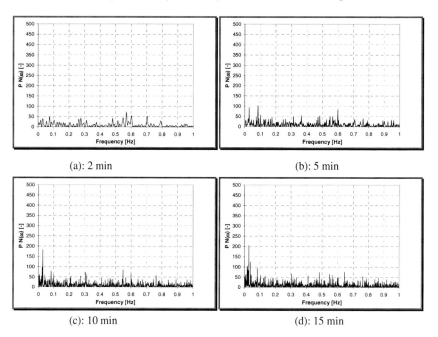

(a): 2 min   (b): 5 min

(c): 10 min   (d): 15 min

Figure 6.21: Comparison of spectra analysis with different measuring times for PBT

6.2.3   Magnitude of MI

6.2.3.1   Introduction

The flow in stirred tanks can be classified into the impeller discharge flow and the bulk flow. In the impeller discharge flow, the instantaneous velocity u' can be decomposed as follows:

$$u' = u'_{BP} + u'_{rand},\tag{6.4}$$

where the subscripts *BP* and *rand* denote the blade passage and pure random or turbulent component. The relative contribution of the regular blade passages to the total fluctuations can be separately estimated in such a way. Otherwise the turbulence fluctuation may be over-estimated by a factor between 4 and 5 (Yianneskis *et al.*, 1993 [146]). Close attention is paid to the discharge flow in the impeller vicinity by performing angle-resolved LDA measurements. The structure and the turbulence properties of trailing vortices, always combined with the blade passage frequency, have been well studied.

In the bulk flow, velocities have been widely treated as pure random or turbulent fluctuations and therefore have stationary features. However, according to the time series analysis of velocities and the dominant macro instability frequency analysis discussed at the beginning of this section, both the bulk flow and the impeller discharge flow are actually superimposed on macro instability motions throughout the whole stirred tank. Hence they deviate widely from the assumed stationary features. The macro instability is the consequence of the precessing vortex core motion in rotational flows, which can be characterized as a good structured oscillation with one or a broad band of dominant frequencies. In a similar way as the treatment on the trailing vortex in the impeller region, the decomposition of stirred flows should involve the macro instability oscillations v'_{MI}. For the impeller discharge flow

$$u' = u'_{MI} + u'_{BP} + u'_{rand},\tag{6.5}$$

and for the bulk flow

$$u' = u'_{MI} + u'_{rand}.\tag{6.6}$$

As already shown in Figure 6.11, in the impeller discharge flow, the velocity fluctuation comprises the fluctuations caused by the individual blade passages, turbulent fluctuations and the fluctuations due to the MI, whereas in the bulk flows, only the last two parts are involved. In order to determine the individual contributions separately, different signal filtering techniques were employed.

6.2.3.2   Decomposition methods

It is obvious that the long-time large-scale flow oscillations due to the MI variation contribute to the evaluation of the RMS velocities. This distortion of the RMS value introduced by the MI variations is very similar to the so called "pseudo-turbulence" phenomenon induced by the individual blade passages, resulting in an overestimation of the turbulence kinetic energy especially in the impeller region. The periodic oscillations induced by the blade passage can be filtered out by applying angle-resolved measuring techniques in which the LDA data are sorted according to the corresponding blade angles. Similar situations appear in the flow analysis influenced by the MI oscillations. In order to analyse the contributions of these long-time oscillations, two main categories of signal processing methods were applied to isolate the low variations from the velocity records in stirred flows with varying degrees of success.

The first category is methods operating in the frequency domain. The most widely applied method is the notch filtering technique described in detail by Press *et al.* (1989) [97]. In combination with the autocorrelation function, a notch filter is used to eliminate the fixed frequencies as well as the coherent frequencies from the raw signal. This type of method was widely applied by Kresta and Wood [67], Rao and Brodkey [101] and Hockey and Nouri [62] in order to distinguish the pure turbulence component from the influence of the blade passages. It should be noted that a notch filter is exclusively appropriate for the removal of a single fixed frequency from the integral of the autocorrelation function, while the oscillation of MI normally comprises a wide range of low frequencies without coherent relations. In this case, the selection of the notch filter window width and the sharpness of the filter function can affect the results of the decomposition significantly. Therefore, notch filter methods are not suitable for the decomposition of the MI frequencies. In the most recent research, wavelets have been introduced into the stirred flow analysis to separate various frequency ranges [86]. The adjustment of the parameters, however, is very complicated and less easily interpreted in the practice, which hinders wider application. For the same objective, Hasal *et al.* [58] employed the proper orthogonal decomposition (POD) technique, allowing reconstruction of the MI oscillation from the POD eigenmodes in combination with spectral analysis. The main disadvantage of this method is the complexity of evaluation and the long calculation time.

The second category for isolating the non-stationary component of the velocity is the smoothing approach working in the time domain. For a variable which is both varying slowly and also embedded with random noise, the noise can be filtered out by replacing each data point by an appropriate local average of surrounding data points without significantly biasing the mean. In the present work, the noise is the pure random turbulent fluctuation, and the long-time variations due to macro instabilities are of most interest. Therefore, in the present study this low-pass filter smoothing technique was employed. Note that during this processing, the selection of the size of the moving averaging window plays an essential role. More details are discussed in Press *et al.* [97].

6.2.3.3 Processing of moving window averaging

Since the velocity signal from LDA is unevenly sampled, the resulting LDA time series is represented by U_i, where $i = 1, 2, \ldots, N$. The corresponding arrival times are t_i, and the intervals between particle arrival times ($\Delta t_i = t_{i+1} - t_i$) are approximately Poisson-distributed. Before the moving averaging, the signal has to be resampled with a fixed sampling rate and transformed into evenly sampled data in order to simplify the later signal processing and to remove any velocity biasing. The raw LDA signal can be interpolated using zero-order interpolation, namely the "sample and hold" interpolation. The value U_i measured at time t_i is retained until the next nearest time interval t_{i+1} unless the raw record jumps to the new value U_{i+1} before this time point. In this way, a new time series is constructed from the raw unevenly sampled data. Once the raw signal has been resampled it is smoothed using a moving average window. Defining that n_L as the number of points used "to the left" of a data point i, i.e., earlier than it, and n_R as the number used to the right, i.e, later, and for fixed $n_L = n_R$, each value U_i is replaced by the average of the data points from U_{i-nL} to U_{i+nL}. This is equivalent to applying a low-pass filter on the time scale of the averaging window. As mentioned above, the raw sampling rate varies depending on the measuring locations in the stirred tank. In order to obtain a uniform signal smoothing quality despite the different sampling rates at different measuring points, either the resampling rate for the raw data or the size of moving averaging window during the smoothing procedure has to be adjusted and predetermined according to the corresponding sampling rate.

Figure 6.22 presents the result of the smoothing procedure for a local point in the middle part of the tank ($r/T = 0.175$, $z/T = 0.575$). The raw time series is shown in Figure 6.22-a, and Figure 6.22-b presents the resampled data with a fixed resampling rate which equals the mean sampling rate in the raw data. No significant deviation due to the biasing during the resampling exists for either the mean or the RMS value. The mean from 10,000 data points changes from 0.1402 to 0.1387 m/s with a deviation below 1.1%, and the RMS value from 0.1574 to 0.1537 m/s with a deviation below 2.3%. The first round smoothing procedure filtered out considerably the small fluctuations in the higher frequency range, namely the pure turbulent fluctuation component. The low-frequency long-time oscillations can be detected as shown in Figure 6.22-c. This type of oscillation is highly periodic and has a fairly organized structure with different levels of amplitude (very similar to Figure 2.15). However, the lowest oscillation is still embedded by higher fluctuations which cannot belong to pure turbulence. Still smaller fluctuations can be removed by a second round of the smoothing procedure as shown in Figure 6.22-d. The shape of the smoothed data after the second round does not differ very much from that after the first round, and the shape of both smoothed data is virtually identical to that of the raw signal in terms of the slow variation on the mean.

The selection of the size of the moving averaging window plays an essential role in the smoothing procedure. Figure 6.23 shows the effect of the average window size on the results of the reconstruction of the long-time low-frequency oscillations. With increasing average

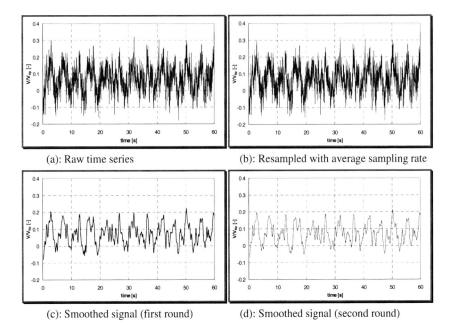

(a): Raw time series

(b): Resampled with average sampling rate

(c): Smoothed signal (first round)

(d): Smoothed signal (second round)

Figure 6.22: Representation of the results during the smoothing procedure

window size, the slow variation becomes smoother. Nevertheless, the smoothed signal cuts off the peaks with low variation, namely the amplitude of low variations is also over-smoothed. Correspondingly, reducing the average window size introduces higher fluctuations overlapping the slow variation. In Figure 6.23-a the averaged sampling rate at points in the bulk flow was about 120 Hz, whereas in Figure 6.23-b in the discharge flow ca. 280 Hz was achieved owing to the higher local velocities. As a consequence, the mean and RMS value of the new reconstructed low variation are likewise strongly sensitive to the selection of the moving smoothing window size. The RMS value decreases sharply with increase in the average window size, whereas the mean value does not vary so considerably. This effect is more evident at points in the bulk flow than in the discharge flow where the periodic fluctuation induced by the blade passage is more pronounced. Figure 6.24 illustrates this effect at a point in the bulk flow.

The time size of the smoothing window was varied with a few different stirrer revolution numbers. By carefully visual determination of an appropriate average window size, the low variations can be isolated from the raw data. In order to eliminate the subjective deviations generated by different visual criteria of different persons or for the same person but at different times, for the flow induced by the RT the time window size was chosen as two revolutions. This value is, however, double the value applied by Roussinova *et al.* [104], who carried out similar evaluations for RT with a similar configuration. This difference is attributed

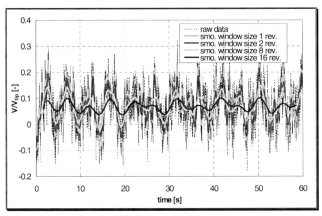

(a): In the bulk flow

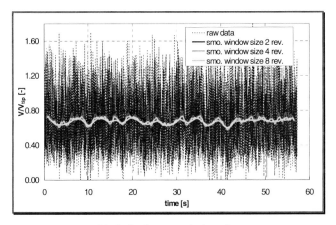

(b): In the flow near the impeller

Figure 6.23: Effect of moving average window size at points in the discharge flow

to the small change during the process. In their work, an adjustment was made already during the resampling process to take into account the different LDA sampling rates at different points, whereas in the present work this factor was only considered during the smoothing. For PBT, since the low variation has a much longer time scale, the time window size was set to eight revolutions.

6.2.3.4 Intensity determination

Once the smoothing process is finished, the pure random turbulence component and the low variations are isolated from each other. One can thereafter estimate the contributions to the

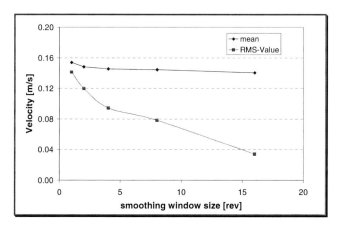

Figure 6.24: Effect of moving average window size on the decomposition

total RMS value of both the pure turbulence component and the long-time low-frequency variations, respectively. The RMS velocity of the low-frequency part can be calculated from the smoothed profiles as

$$u'_{MI} = \sqrt{\frac{1}{n}\sum_{i=1}^{n}(U_{MI} - \bar{U})^2} \, , \qquad (6.7)$$

where u'_{MI} represents the RMS velocity of the low variation due to the MI, U_{MI} the instantaneous velocity in the smoothed value and $\bar{U}$ the mean value of the low variation profile. The intensity of the RMS velocity due to the MI can be then characterized as

$$\text{Intensity}_{MI} = \frac{u'_{MI}}{u'} \, , \qquad (6.8)$$

where u' is the RMS velocity before the smoothing procedure. Subsequently, the contribution of the pure turbulence component can be quantified by the following equation:

$$u'_{turb} = \sqrt{\frac{\sum_{i=1}^{n}(U - U_{MI})^2}{n-1}} \, , \qquad (6.9)$$

where U symbolizes the original instantaneous velocity at each moment and u'_{turb} represents the pure turbulence component portion of the total RMS velocity. The intensity of the RMS velocity due to the pure turbulence component can be deduced similarly as

$$\text{Intensity}_{turb} = \frac{u'_{turb}}{u'} \, . \qquad (6.10)$$

Ultimately, the contribution of the macro instability motions to the turbulence kinetic energy can be estimated if all three RMS velocity components are considered:

$$k_{MI} = \frac{1}{2}(u'^{2}_{MI} + v'^{2}_{MI} + w'^{2}_{MI}) \, . \tag{6.11}$$

The pure turbulence kinetic energy can be calculated in the same way:

$$k_{turb} = \frac{1}{2}(u'^{2}_{turb} + v'^{2}_{turb} + w'^{2}_{turb}) \, . \tag{6.12}$$

This evaluation procedure was, for example, carried out for velocity time series at point $r/T = 0.175$, $z/T = 0.575$ in the bulk flow which was shown in Figure 6.23-a. The RMS value due to the low variation is 0.1198 m/s. Thus the intensity of the MI can be calculated according to Equation (6.8) as 76%. Embedded on this low variation, the pure turbulence component can be evaluated according to Equation (6.9); this value is 0.0866 m/s and has a portion of 55% of the total RMS value of 0.1574 m/s. It should be pointed out that the total RMS value u' is not a simple sum of the low variation part u'_{MI} and the pure turbulent portion u'_{turb}.

6.2.3.5   Spatial distribution of the magnitude of the low variations

RT

Based on the evaluation procedure mentioned above, the magnitude of the low variations can be compared at different positions in the tank. For the standard RT configuration, the radial component was selected for comparisons

In the discharge flow (profile $z/T = 0.325$), as depicted in Figure 6.25-a, the total RMS velocities are the highest near the blade, and in this region the so-called "pseudo-turbulence" due to the blade passages dominates. In comparison with the relatively flat shape of the RMS velocity profile due to the low variations, the total RMS-velocity profile has a very similar shape to that of the RMS velocity induced by the pure turbulence component, indicating that the flow at this vertical level is more influenced by the blade passing effects than the low variations. The intensity of the turbulence component achieves more than 90% in the whole profile, whereas the MI intensity is below 50% at most points. Figures 6.24-b and c present the magnitude analysis in the bulk flow below and above the stirrer, respectively. With increasing distance from the stirrer blades, the intensity of the pure turbulence component decreases. The contribution of the low variations becomes more and more pronounced. This can also be proved by the more similar profile shape between the total RMS velocity and that of MI. Up to the bulk flow near the free surface shown in Figure 6.25-d, it is obvious that the flow is dominated by the MI movements. The intensity of MI exceeds 80% in the region near the shaft and about 70% in the middle part.

As a common feature in all flow regions, the magnitude of the MI part in the side near the geometric centre is always higher than in the side near the tank wall, indicating a stronger influence of the low variations near the flow rotation centre due to precessing vortex core motion. The magnitude of the low variations at all positions is around a constant level of 0.05 V_{tip}, whereas the pure turbulence component damps off with increasing distance from the stirrer which is acting as the turbulence source. This feature confirms again that the low variations movements of the MI are large-scale variations having the order of the tank size.

A similar comparison was also carried out between different velocity components. It can be clearly seen from Figure 6.26 that the two horizontal components, namely the radial (Figure 6.26-a) and the tangential (Figure 6.26-b) components, have clearly higher RMS velocities due to the MI motion than the axial component (Figure 6.26-c). This difference between the horizontal components and the axial component is remarkable in the region near the shaft ($r/T < 0.2$), in which also a higher spectral power exists in the frequency spectra. This more intensive MI motion causes an increase in the total RMS velocities. Note that the total RMS velocity involving the MI contribution represents a strongly anisotropic turbulence, since the axial RMS velocity is much smaller than the other two. This anisotropy is actually induced by the MI motion. The pure turbulent portion of all three components stays at a fairly constant level (ca. 0.03 V_{tip}), indicating the existence of isotropy of the turbulence even in the flow near the surface, if the influence of the low variation motion is filtered out by the decomposition techniques. In addition, the pure turbulence is homogeneous along the radial direction, and the turbulence kinetic energy has an order of 0.012 V_{tip}^2. This conclusion is of importance for CFD application in stirred flows. The widely applied k-ε model is known to be inappropriate in rotating and/or highly three-dimensional flows. Derksen et al. (1999) [31] demonstrated using an invariant map that the turbulence in several flow regions close to the impeller is anisotropic. Similar results were also obtained by Schäfer et al. [110] from their angle-resolved LDA data, and they pointed out that only in the direct vicinity of the impeller blade does anisotropy of turbulence exist. In other regions the flow is considered fully isotropic, and therefore it is still meaningful to use the k-ε model. The present work confirmed this assumption. Moreover, the higher turbulence kinetic energy existing in experimental investigations (Schäfer et al., 1999) compared with the numerical work can be explained by the additional contribution of the MI motions which are not taken into account in numerical investigations for stirred flows.

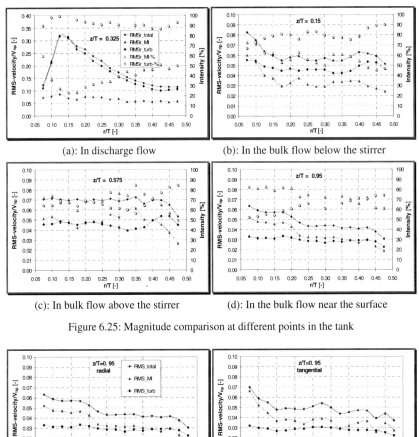

(a): In discharge flow

(b): In the bulk flow below the stirrer

(c): In bulk flow above the stirrer

(d): In the bulk flow near the surface

Figure 6.25: Magnitude comparison at different points in the tank

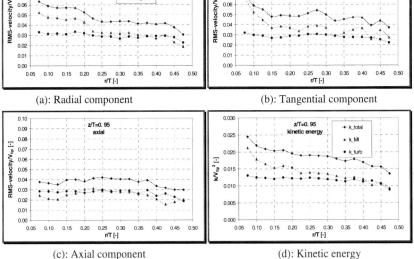

(a): Radial component

(b): Tangential component

(c): Axial component

(d): Kinetic energy

Figure 6.26: Magnitude comparison between different components

PBT

For PBT, similar tendencies occur in the flow near the surface as for RT. A profile at the same vertical level as for RT was selected and the results are described in Figure 6.27. Also for PBT, the contributions to the formation of the total RMS value from both horizontal components are much higher than from the axial component, and this difference is greater than that of RT. As shown in Figure 6.27-c, the contribution of the MI low variations is very weak in comparison with that of the pure turbulence. The portion of the MI low variations in the formation of kinetic energy is higher than that of the pure turbulence component in the side near the shaft, while the pure turbulence component contributes more to the turbulence kinetic energy in the side close to the wall. As for RT, the pure turbulence again shows the features of homogeneity and isotropy. The magnitude of the pure turbulence has the same level as that of RT ($0.01\ V_{tip}^{2}$). Note that a much longer smoothing window time was applied in the decomposition procedure for PBT. This indicates that at the surface the flow induced by a PBT is much more dependent on the MI motions than that by RT. Actually, the macro instability motions are the main mixing mechanisms in the uppermost part of the stirred tank.

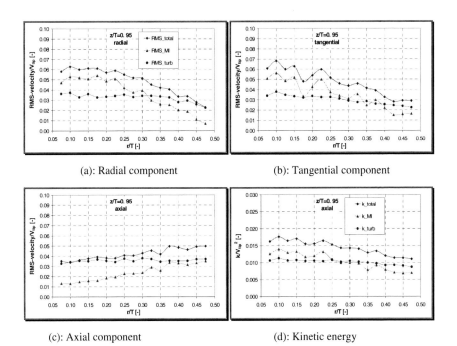

(a): Radial component   (b): Tangential component

(c): Axial component   (d): Kinetic energy

Figure 6.27: Magnitude comparison between different components for the PBT

6.2.3.6   Dependency of the MI magnitude on rotational speeds

The magnitude of the MI motions was also investigated for different impeller rotational speeds for RT. A profile near the surface and the shaft ($z/T = 0.95$ and $r/T = 0.08$-0.2) was selected for detailed analysis since the flow at the surface is of great interest in the present work. The rotational speed of the impeller was varied from 250 to 375 rpm, corresponding to a Reynolds number range from 4520 up to 6780. The k_{MI} and the k_{turb} values are normalized by V_{tip}^2. The results are summarized in Figure 6.28.

As shown in Figure 6.28-a, the dimensionless turbulence kinetic energy of the MI portion increases with the Reynolds number. The dimensionless kinetic energy part of MI motions increases with the Reynolds number and decreases with the radial distance from the shaft. The MI motion cannot be scaled by the rotational speed. In contrast, as shown in Figure 6.28-b, a nearly constant dimensionless kinetic energy part of the pure turbulence occurs in the range of

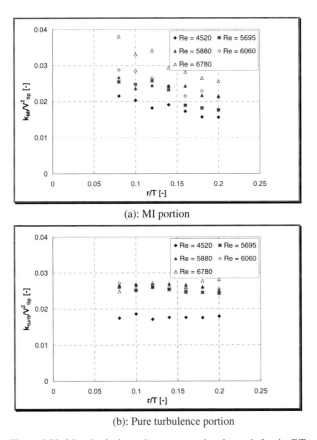

(a): MI portion

(b): Pure turbulence portion

Figure 6.28: Magnitude dependence on rotational speeds for the RT

$0.027\, V_{tip}^2$ provided that the Reynolds number is higher than 4520. This indicates that the stirred flow is still turbulent even near the surface.

Test case	D/T [-]	C/T [-]		Blade No.	D [mm]	D_{Disk} [mm]	L_{Blade} [mm]	W_{Blade} [mm]	T_{Bldade} [mm]	N [rpm]
Task 1		**Blade number variation**								
4 baffles	0.33	0.33		2	132	99	33	26.4	2.0	300
4 baffles	0.33	0.33		3	132	99	33	26.4	2.0	300
4 baffles	0.33	0.33		4	132	99	33	26.4	2.0	300
4 baffles	0.33	0.33		6	132	99	33	26.4	2.0	300
4 baffles[1]	0.33	0.33		0	99	99	-	-	2.0	533
Task 2		**Baffle number variation**								
Unbaffled	0.33	0.33		6	132	99	33	26.4	5.0	300
2 baffles	0.33	0.33		6	132	99	33	26.4	5.0	250
4 baffles	0.33	0.33		6	132	99	33	26.4	5.0	300
Task 3		**Clearance and filling height variation**								
4 baffles	0.33	0.33		6	132	99	33	26.4	5.0	300
4 baffles	0.33	0.4		6	132	99	33	26.4	5.0	300
4 baffles	0.33	0.5		6	132	99	33	26.4	5.0	250
4 baffles	0.33	0.33	0.89^2	6	132	99	33	26.4	5.0	250
4 baffles	0.33	0.33	0.75	6	132	99	33	26.4	5.0	250
Task 4		**D/T ratio variation**								
4 baffles	0.225	0.33		6	90	67.5	22.5	18	2.0	613
4 baffles	0.33	0.33		6	132	99	33	26.4	2.0	300
4 baffles	0.45	0.33		6	180	135	45	36	2.0	150
4 baffles	0.65	0.33		6	260	195	65	52	2.0	85

Table 6.3: Geometry and operating conditions applied in the variation investigations

[1] A disk with a diameter of 99 mm and the rotational speed was set to keep a constant Reynolds number.

[2] The H/T value

6.2.4   Investigations of variations in Macro instabilities

6.2.4.1   Introduction

In order to investigate the impacts of the geometry configurations on the characteristics of the dominant frequencies, a series of variations in geometry configurations were carried out systematically. The variations were based on the standard configuration of RT, correspondingly, varying in blade numbers, baffle numbers, clearance, liquid filling height, D/T ratios etc. The rotating disk is also brought into the comparison with the RT. Table 6.3 gives a detail summarization of the geometry variations. For the convenience of comparisons, a horizontal profile near the surface ($z/T = 0.95$) for all cases was selected. The dominant frequencies were evaluated from the radial velocity component for all variations, and the angular velocity was also involved in comparisons, in order to analyse the influence of the rotational flow on the MI motions.

6.2.4.2   Results and discussion

Blade number investigation

For comparison, the tangential component in a horizontal profile near the surface ($z/T = 0.95$) was measured for each RT, the rotating disk and PBT. The investigation results are summarized in Figure 6.29-a. For RTs, with increase in the blade number, the MI dominant frequencies increase gradually. Note that the MI dominant frequencies of the rotating disk and of the PBT are much smaller than those of the RTs. The dominant frequencies scatter in an inverse order. The dominant frequencies scatter more strongly if the blade number decreases. More detailed analysis for two blades and three blades shows that instead of one peak, two or three harmonic peaks appear in the power spectrum, leading to a leap in the dominant frequencies and thus scattering of the dominant frequencies. This bifurcation occurs more often in the range $r/T > 0.25$, corresponding to weaker rotational flow regions depicted in Figure 6.29-b. In this region, the flow is influenced by different rotational speed ranges, resulting in the appearance of different dominant frequencies both in time and in space.

Figure 6.29-b presents comparative results for the rotational speed at the corresponding radial positions n_r. The local flow rotational speed n_r is normalized by the impeller speed N for easier comparison. For RTs, all angular velocity profiles are similar, representing a similar rotational flow pattern, but with slight differences in magnitude. The rotational speeds increase gradually with increase in the blade number, leading to an increase in the dominant frequency. Despite the much smaller magnitude, the rotational speed of the disk displays a similar profile shape to RTs, whereas the dimensionless rotational speed of PBT shows a totally different profile. The magnitude has a relative flat level around 0.02 along the whole radial profile,

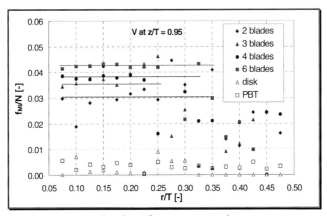

(a): Dominant frequency comparison

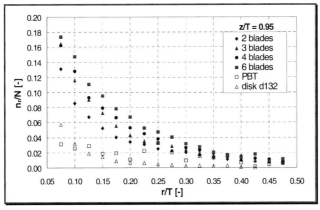

(b): Rotational speed comparison

Figure 6.29: Comparison of blade number variations

indicating a weak rotational flow. This different profile from that of RTs explains the greater scattering features of the MI dominant frequencies.

The angular velocity of the flow influences the magnitude of the dominant frequencies provided that they have a similar profile. This conclusion can be better confirmed by Figure 6.30. The tangential velocity component at the same point for all RTs and the rotating disk are plotted against the corresponding dominant frequency. Reasonable linearity between the tangential velocity and the dominant frequency can be seen. Based on the similar flow pattern, the dominant frequency depends directly on the rotational speed of the flow.

The magnitudes of the MI motions for all RTs are also compared, and the results are illustrated in Figure 6.31. The intensity of the low variations for 3-6 blades increases slightly with

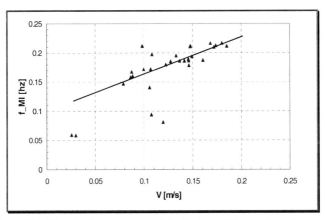

Figure 6.30: Linearity between tangential velocity and the dominant frequency

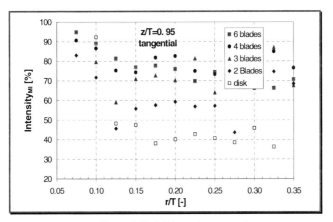

Figure 6.31: MI intensity comparison

increasing blade number. However, for two blades this intensity decreases considerably, and the intensity for the rotating disk has the smallest value. In the last two cases the precessing vortex core motion is very weak.

Baffle number variations of RTs

The results of the dominant frequency in the tank with different baffle numbers are summarized in Figure 6.32-a. No difference can be observed between the four- and two-baffle configurations, although considerable differences exist between the flow rotational speeds shown in Figure 6.32-b. With more baffles, the rotation is disturbed more strongly and thus the free vortex structure of the rotation is more suppressed.

For the unbaffled configuration, the dominant frequency of the radial component scatters totally. No uniform dominant frequency can be found. Moreover, the dominant frequency of the tangential velocity component jumps to a much higher level by a factor of ca. 10. The reason for this scattering can be explained by Figure 6.33. Despite the much higher rotation rate for the unbaffled configuration, no dominant frequency exists, indicating a very weak precessing vortex core motion in unbaffled tanks. The other geometry imperfections do not introduce strong enough disturbances. This result confirms the visual observations in Section 5.1.2 that the disturbance of baffles on the flow rotation induces precessing of the rotation vortex core.

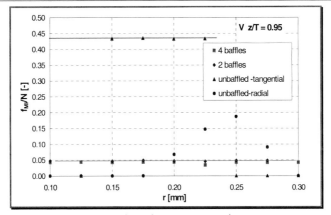

(a): Dominant frequency comparison

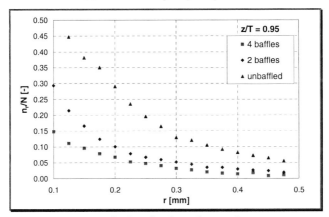

(b): Rotational speed comparison

Figure 6.32: Comparison of baffle number variations

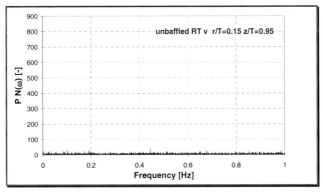

Figure 6.33: Power spectral examples of unbaffled configuration

Clearance and filling height variations of RT

Three clearances, $C/T = 0.33$, 0.4 and 0.5, were investigated. Comparative results are depicted in Figure 6.34-a. For $C/T = 0.33$ and 0.4, the dominant frequencies are very close to each other (f_{MI}/N in the region of 0.042). This shows that the clearance does not affect the dominant frequency. Nevertheless, as the C/T ratio is further increased to 0.5, the dominant frequency jumps suddenly from 0.042 to 0.11, even though the rotation rates for all three cases are in a very close range (Figure 6.34-b).

The dependence of the dominant macro instability frequencies on the bottom clearance has attracted considerable attention in investigations in the literature. Nevertheless, no convincing

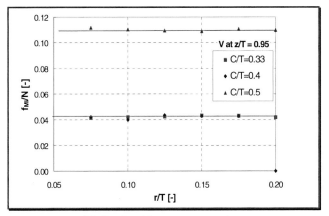

(a): Dominant frequency comparison

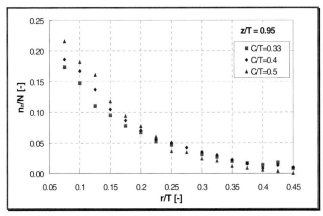

(b): Rotational speed comparison

Figure 6.34: Comparison of clearance variations

explanation has been obtained yet. Since the macro instability is a long-time phenomenon having the time scale of the turnover time of circulations, Bruha et al. [16] tried to associate the dominant macro instability frequencies of axial impellers with the primary circulation time which depends on the pumping capacity of the impeller. The circulating time θ_c in a tank having a volume V can be estimated from the following equation:

$$\theta_c = \frac{V}{Q} = \frac{V}{Fl \cdot ND^3},$$
(6.13)

where Q represents the pumping volume flow rate and Fl the primary flow number character-izing the pumping capacity. The primary flow number is, in turbulent flow regimes, inde-pendent of the Reynolds number. Equation (6.13) can be rewritten for the case $H = T$ as

$$f_c \equiv \frac{1}{N\theta_c} = \frac{4}{\pi} fl(D/T)^3,$$
(6.14)

where f_c characterizes the mean frequency of primary circulation, and is expected to have a certain relation with the dimensionless MI frequency f_{MI}^*. Bruha et al. succeeded in confirm-ing this correlation for the flow induced by a PBT with the bottom clearance in the range $C/T = 0.2$-0.5.

For this purpose in the three C/T configurations, angle-resolved LDV measurements near the blade were carried out in the present study. The primary flow number was calculated accord-ing to Equation (2.33), and the primary volume flow rate was calculated by

$$Q = 6 \left(\frac{2\pi r}{360} \right) \int_{z_1 = C-h/2}^{z_2 = C+h/2} \int_{\phi=0°}^{\phi=60°} \bar{U}\left(r = D/2, \phi, z \right) \, \mathrm{d}\phi \mathrm{d}z,$$
(6.15)

where C represents the clearance of the impeller, h the blade height and ϕ the impeller blade angle. The results are summarized in Table 6.4. The results presented here cannot confirm the assumption mentioned above for the RT. The primary flow number Fl increases gradually with increasing impeller bottom clearance (0.58 up to 0.63). The f_{MI}^* values in the first two cases are nearly identical. For the case $C/T = 0.5$, despite having approximately the same flow

C/T [-]	Fl [-]	f_c [-]	f_{MI}^* [-]
0.33	0.58	0.027	0.042
0.4	0.59	0.027	0.041
0.5	0.62	0.028	0.108

Table 6.4: Macro instability dominant frequency and the primary flow number

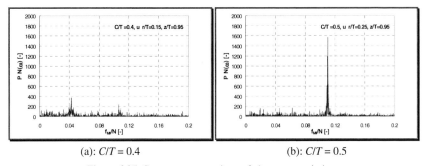

(a): $C/T = 0.4$ (b): $C/T = 0.5$

Figure 6.35: Spectrum comparison of clearance variations

number, the MI frequency is much larger in the first two cases.

The reason for this leap in the dominant frequency can be better understood by detailed analysis of the power spectra for the corresponding cases. Two spectra corresponding to the last two cases are presented in Figure 6.35-a and b. For $C/T = 0.4$, accompanying the first peak which is in the same range as that of $C/T = 0.33$, a second peak in the range around 0.105 appears in the spectrum with a relatively higher spectral power level in comparison with the first peak. Likewise, the power spectrum for the case $C/T = 0.5$, which corresponds to an impeller submergence depth $E/T = 0.5$, has a similar power distribution. In the lower frequency range, a few peaks appear in the range close to that of $C/T = 0.33$. The most remarkable feature in the spectrum is the peak near 0.11 with a much higher magnitude (at most points above 1,500), indicating a more pronounced resonance frequency. Compared with this, for the case $C/T = 0.4$, the spectral power of all peaks is clearly smaller than that for $C/T = 0.33$, indicating a transition case in which both peaks are suppressed.

Similar tendencies can also be found as the filling height is varied from $H/T = 1$ to 0.75. As the filling height is decreased to $H/T = 0.89$, corresponding to an an impeller emerging depth $E/T = 0.56$, the dominant frequency appears near 0.11 which is shown in Figure 6.36-a. This value shows good agreement with the case $C/T = 0.5$, which has a similar impeller submer-

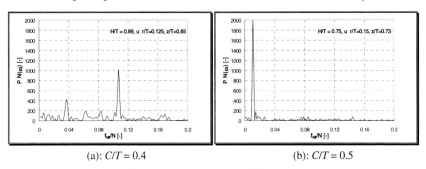

(a): $C/T = 0.4$ (b): $C/T = 0.5$

Figure 6.36: Spectrum comparison of filling height variations

gence emerging depth ($E/T \approx 0.5$). As the H/T ratio is further decreased to 0.75 ($E/T = 0.42$), the dominant frequency drops considerably to 0.01 as shown in Figure 6.36-b. In this case, the impeller is again located nearly in the middle between the surface and the bottom as in the case $C/T = 0.5$, and correspondingly a much more pronounced dominant frequency (spectral power close to 2000) appears

Based on the result of the above-discussed submergence variations, it can be concluded that the dominant frequencies are distributed into a few levels (e.g. 0.01, 0.04, 0.11), and these frequencies are sensitive to the impeller submergence depth and the height ratio between the upper part and lower part separated by the impeller. The submergence depth can be associated with the aspect ratio L in the Ekman layer. For example, for $C/T = 0.33$, the upper part has an aspect ratio of about 1.33, whereas the lower part has only a half of this value. An inherent relationship between different dominant frequencies was not found in the present work.

With variation of the submergence depth of the impeller, these levels of dominant frequencies variiy in magnitude, and the dominant frequency leaps between these levels. As the geometry is changed to a situation under which the resonance conditions for a certain frequency are better satisfied, a low variation with this frequency will mostly be stimulated. Among them, the case $C/T = 0.5$ was investigated in detail on the flow pattern, as shown in from Figure 6.37. Strict symmetry exists between the upper and lower parts of the tank both in circulation pattern and in angular velocity. Therefore, the two rotational flows together stimulate a resonant

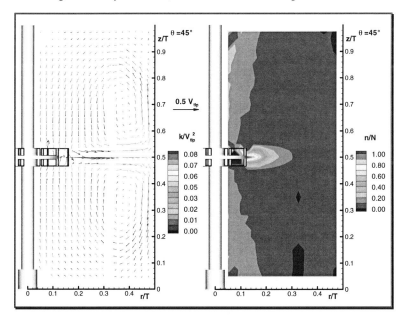

Figure 6.37: Mean flow the RT with $C/T = 0.5$ in the r-z plane $\theta = 45°$ in the fully baffled tank

frequency, and in this case the macro instability motions have a high variation amplitude.

Diameter ratio variation investigations

The influences of the D/T ratio on the dominant frequency were investigated for four RTs with different diameters, namely 90, 133, 180 and 260 mm, corresponding to D/T ratios of 0.225, 0.33, 0.45 and 0.65. The dominant frequencies are depicted in Figure 6.38-a. For smaller D/T ratios ($D/T = 0.225$ and 0.33), the dominant frequency is uniformly distributed. If the ratio increases to 0.45 and 0.65, the dominant frequencies scatter strongly, and the magnitude in power spectrum decreases considerably. The rotation rate of the flow increases strongly, however, with increasing D/T ratio. For $D/T = 0.65$, the rotation profile has a similar

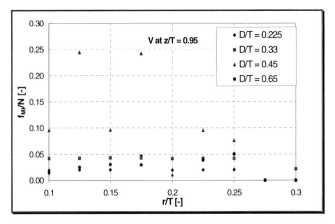

(a): Dominant frequency comparison

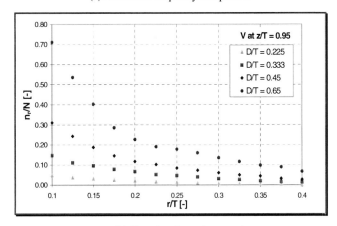

(b): Rotational speed comparison

Figure 6.38: Comparison of diameter variations

shape to that of the unbaffled configuration, indicating a status of unbaffled configuration. This can be attributed to the fact that the influence of baffles is to some extent overcome. Accompanying this, the disturbance effect is likewise decreased. Thus, the precessing vortex core motion is suppressed.

Brief conclusions

Based on the discussion mentioned above, the blade number investigation shows that if the geometry and flow pattern are very similar, the dominant frequency depends linearly on the rotation rate of the flow. The baffle disturbs the rotational flow, and thereafter induces the precessing vortex core motion. The most important factors influencing the dominant frequency are the submergence depth of the impeller and the symmetry of the flow pattern between the upper and lower parts in the tank. The diameter does not affect the dominant frequency significantly provided that the baffle disturbance effect is not overcome.

6.3   Wall Jet Flow Before Baffles

6.3.1   Introduction

For RT, the surface vortex is strongly disturbed and destabilized by the swelling axial surges before the baffles, which vary in intensity with time. For PBT, the dominant direction of non-rotational flow at the surface is directly induced by varying swelling surges before the baffles. This dominant flow direction generates surface vortices according to the visual observations. Therefore, these axial surges before the baffles have a significant impact on air entrainment at the free surface.

This flow is generated in the plane of the impeller by reorientation of the baffle from radial and tangential directions to an axial direction. By impinging and directing at the vessel wall, this flow extends upwards along the vessel wall, but decreases in intensity with increasing height. This is similar to a three-dimensional wall jet flow. Turbulent scaling for wall jets has been widely examined (Glauert, 1956 [51]; Padmanabham and Gowda, 1991 [91]). Glauert divided the wall jet flow regime into an inner layer, between the wall and the points of maximum velocity, and an

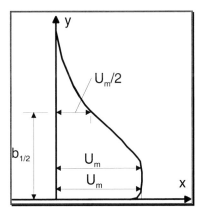

Figure 6.39: Schematic representation of wall jet flow field

outer layer. He matched the solutions for these two layers at the point of maximum velocity of the profile where the shear stress was assumed to be zero. Therefore turbulent wall jet flows can be scaled, as shown in Figure 6.39, by the characteristic velocity and length scales, namely the local mean maximum velocity u_m and the half-width of the jet $b_{1/2}$. Here the half-width represents the point at which $u/u_m = 0.5$ on the dimensionless velocity profile, and the distance from the wall is normalized by dividing by the half-width. When the wall jet flow remains fully turbulent, the dimensionless profile retains similarity at different streamwise positions. Bittorf and Kresta (2000) [9] characterized such a kind of wall jet flow in a stirred tank flow induced by a pitched blade turbine, and further examined the limits of fully developed turbulence in the bulk of a stirred tank. Conclusions were drawn that the upper third of the tank drops into the transitional flow regime even at $Re = 2 \times 10^4$ regardless of different C/T ratios.

In the steady flow pattern such a wall jet can also be found for both RT and PBT. Instead of arising from the bottom for a PBT, the wall jet flow for a Rushton turbine is formed directly on the wall by the diverting action of the baffles in the impeller plane. For the purpose of the wall jet flow analysis, the axial velocity component in the plane $\theta = 5°$ was measured by LDA for the RT and the PBT. The geometry and operating conditions were given in Table 6.1.

6.3.2   Results of mean velocity profiles

Since the flow below the height $z/T = 0.47$ is still strongly influenced by the upper circulation loop (see Figure 6.4), the wall jet flow profile cannot be scaled correctly. Therefore, the profiles were selected above the height $z/T = 0.475$, corresponding to the upper vortex ring centre. In the first step, only the profiles at the lower heights ($z/T = 0.475\text{-}0.575$) were taken to check the similarity of the wall jet flow before the baffles. As shown in Figure 6.40-a, similarity is satisfied at different vertical heights, indicating that the wall jet flow before the baffles in this region is fully turbulent. For further analysis, these profiles were taken for the regression characterizing this type of turbulent wall jet flow. A cubic regression is required to account for both the inflection point and the maximum at $u/u_m = 1$. The resulting cubic regression can be written as

$$u/u_m = 0.101\eta^3 - 0.498\eta^2 - 0.136\eta + 1.026. \qquad (6.16)$$

This cubic regression has a correlation coefficient $R^2 = 0.99$. This regression is valid from the maximum velocity to the point at which the dimensionless velocity passes through zero. Beyond this point, the wall jet is no longer examined for similarity because it is affected by the impeller discharge flow.

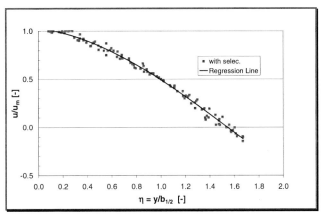

(a): Regression line

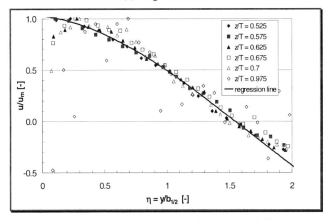

(b): Vertical height effects on the similarity

Figure 6.40: Wall jet flow similarity of profiles at different heights for the RT

The regression result can be further used to examine the similarity for jet flows at different heights. Figure 6.40-b shows that the wall jet profiles can retain similarity only up to $z/T = 0.675$. Profiles with axial positions higher than $z/T = 0.675$ tend to be unable to retain similarity. The standard deviation from the regression line σ jumps significantly from about 0.02 at $z/T = 0.675$ to about 0.07, when the height of the profiles exceeds $z/T = 0.7$. This indicates that at such a Reynolds number ($Re = 5424$), the axial wall jet flow can only remain turbulent up to a height $z/T = 0.675$, despite the fact that the mean axial velocity achieving 0.05 V_{tip} extends up to a height $z/T = 0.93$. At that height the wall jet flow become transitional. The flow near the surface is not fully turbulent, and this is in good agreement with the theory of Liepe *et al.* [73].

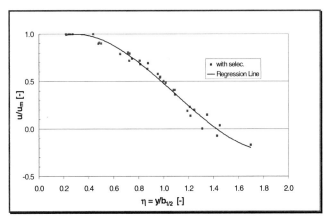

(a): Regression line

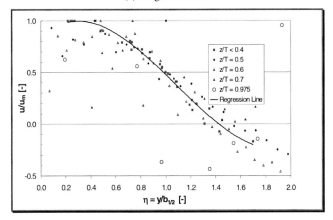

(b): Vertical height effects on the similarity

Figure 6.41: Wall jet flow similarity of profile at different heights for the PBT

For PBT, the regression line was taken from the height range $z/T = 0.05$ to 0.4. The result is presented in Figure 6.41-a. In comparison with the RT, the correlation coefficient has a lower value ($R^2 = 0.985$), indicating the greater complexity of the wall jet flow induced by the PBT due to the bottom redirection effects. The resulting cubic regression is expressed as

$$u/u_m = 0.6476\eta^3 - 2.101\eta^2 + 1.0878\eta + 0.8422 . \qquad (6.17)$$

This regression is very similar to the result of Bittorf and Kresta [9] for a PBT in a similar geometry.

As shown in Figure 6.41-b, the dimensionless wall jet flow shows substantial deviations from the similarity as the height reached $z/T = 0.5$ (with a standard deviation σ of 0.2136). The standard deviation increases strongly with increasing height (at $z/T = 0.6$ with $\sigma = 0.233$ up to 0.297 at $z/T = 0.7$). The wall jet flow is no longer fully turbulent. At the surface at $z/T = 0.975$, the dimensionless profile scatters strongly, indicating the non-turbulent flow regime at the surface.

6.3.3   Results of semi-instantaneous velocity profiles

The results from the mean velocity profiles cannot agree well with the strong swelling upward surges before the baffles at some instants described in the visual observations. It was shown by the visual observations and the macro instability analysis that the flow varies its pattern with a common rhythm. That means that at some instants, the wall jet flow before one baffle achieves a much higher velocity at the same time, and then at another instant a much lower velocity, leading to the varying surges in location before different baffles.

Based on this assumption, the velocity time series at different points before the baffle were analysed. If one velocity time series of a single point is sorted according to the magnitude of the momentary velocity, and thereafter the velocities are normalized by the maximum velocity at this point into a dimensionless value, a direct comparison between different spatial points can be carried out. As shown in Figure 6.42, at different measuring points (point $r/T = 0.25$, $z/T = 0.95$ and point $r/T = 0.45$, $z/T = 0.95$) the velocities have the similar distribution forms. On both the minimum and maximum sides, the velocity decreases and increases sharply, whereas in the middle part the velocity varies approximately linearly. The sharp form on both sides can be expected as a consequence of the macro instability motions. For approximately constant sampling times, only the velocities in the sharp range (about 150 points) are taken for averaging. In such a way, semi-instantaneous wall jet profiles can be approximately approached.

The semi-instantaneous wall jet profile is compared with the mean velocity profile at the same height $z/T = 0.975$ which is very close to the surface. Comparative results for the RT and the PBT are illustrated in Figure 6.43-a and b, respectively. For both the RT and the PBT, compared with the mean velocity profile, greater similarity is

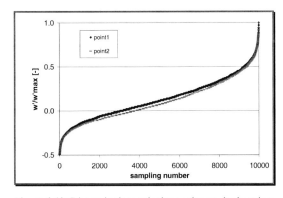

Figure 6.42: Dimensionless velocity sorting at single points

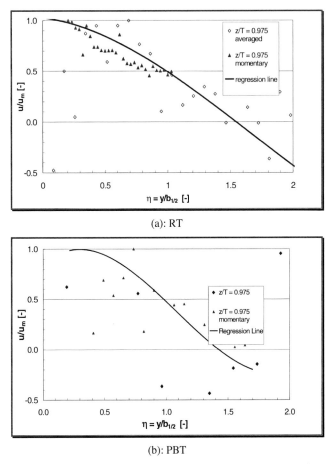

(a): RT

(b): PBT

Figure 6.43: Similarity comparison between mean and instantaneous velocity profiles

achieved for the momentary jet flow even at a height near the surface. The standard deviation decreases considerably from 0.1 to 0.07 for the RT and from 0.786 to 0.428 for the PBT. Still weakly turbulent momentary wall jet flow occurs at the surface for the PBT. However, the profile is closer to the regression line than that of the mean.

This comparison indicates that the wall jet flow along the baffles can remain more turbulent at certain moments than the mean flow pattern. The macro instability motions intensify the wall jet flow with a common rhythm. Strongly turbulent wall jet flow can also be confirmed by the instantaneous flow field shown in Figure 6.44, which was visualized in two different vertical planes using the laser sheet visualization technique in a geometrically similar configuration, but at a much higher Reynolds number ($Re = 49,000$). Compared with the irregular flow pattern in the uppermost part close to the vessel wall in plane 45°, in plane 5° the instantaneous

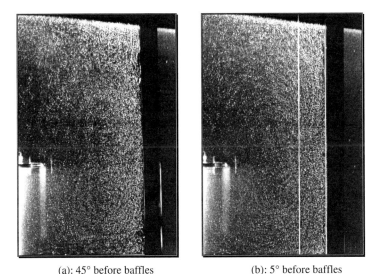

(a): 45° before baffles   (b): 5° before baffles

Figure 6.44: Instantaneous flow field in different vertical planes

wall jet flow before the baffles remains a strongly axial velocity up to the surface with little difference in magnitude.

6.4 Improvement of Avoiding Surface Aeration

The present work was initiated with the objective of avoiding surface aeration in stirred vessels. From this point of view, two improved possibilities for avoiding surface aeration were developed based on the investigations and further verified in the present work. Both possibilities were realized first in stirred vessels filled with water in order to determine the critical rotational speeds for air entrainment by visual observations. The flow at the surface of the stirred vessel was then quantitatively measured by LDA in the silicone oil mixture.

6.4.1 Shaft baffles

As introduced in Sections 2.1.6 and 5.1.1, the free liquid surface will follow the pressure distribution if the fluid rotates. For unbaffled tanks, a central vortex having a Rankie vortex structure appears. Wall baffles can suppress the formation of this central vortex by disturbing the angular velocity effectively. Based on the results of visual observations at the free surface, it was noted that in the uppermost region close to the shaft, the fluid is still in a strong rotational motion, indicating the invalidity of the wall baffles in this region. For radial impellers, e.g. RT, the surface vortex appears which manifests the precessing vortex core of this strongly rotational flow. In a similar way, it is expected that the strong rotational motion can be sup-

pressed likewise by introduc-
ing baffles mounted close to
the shaft, and thereafter eliminate the formation of surface
vortices. This type of baffle is
termed here the shaft baffle.

Figure 6.45 illustrates schematically the geometry of
shaft baffles. Four radial baffles were mounted close to the
shaft symmetrically on the
circumference. The baffles are
shifted by 90° to the wall baffles. The shaft baffle has a
width $W_{baffle} = 30$ mm and a

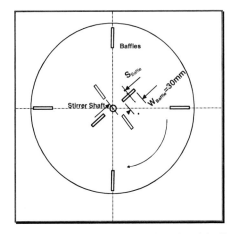

Figure 6.45: Schematic representation of shaft baffles

distance to the shaft $S_{baffle} = 20$ mm. The thickness of the baffle is 5 mm. The submergence
depth of the baffles was set to 80 mm.

The critical rotational speeds were determined by visual observations. N_{cloudy} was increased
from 264.0 to 293.0 rpm, i.e. by about 11%. However, the first bubble speed N_{CSA1} was even
20% lower than that of the standard configuration. This is attributed to the appearance of the
tail vortex behind the shaft baffles; this type of tail vortex occurs very occasionally, and thus
cannot put air bubbles in the liquid continuously. The rotational motion of the liquid is
strongly disturbed, the surface vortex is effectively suppressed.

The LDA measurement explains the validity of the suppressing effect of shaft baffles on air
entrainment. The flow was compared with the standard configuration at the same points
($z/T = 0.95$, $r/T = 0.08$-0.2) both in mean values and in macro instability features. As shown in
Figure 6.46-a, in comparison with that of the standard configuration, the tangential velocity
component has been substantially decreased in flow with shaft baffles. The RMS values of the
tangential component remain at the same level, even with a slightly lower magnitude, in contradiction to the obstacle effects of baffles which normally generate more turbulence. Figure
6.46-b presents the comparative results of turbulence kinetic energy. The shaft baffles have
weakened the turbulence kinetic energy significantly. This is realized by the decrease in both
the MI contribution and the pure turbulence contribution. The MI dominant frequency has not
changed except at points close to the shaft baffles, indicating that the shaft baffles cannot
eliminate the precession vortex motion.

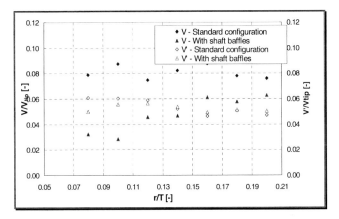

(a): Tangential velocity component

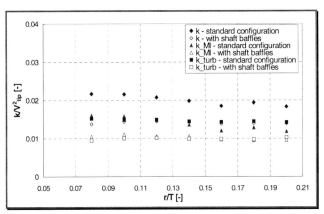

(b): Turbulence kinetic energy

Figure 6.46: Comparison between flow with and without shaft baffles

6.4.2   Surface screens

The second possibility is the application of a screen mounted horizontally near the surface. It is well known that a screen can either intensify or weaken turbulence depending on the geometry [61]. It is widely applied to generate homogeneous turbulence, e.g. in a wind tunnel. The most import geometry feature of a screen is the ratio l/d, where l denotes the grid space and d the screen grid width. In addition, Ramanurthi and Tharakan (1996) [100] showed in their flow visualisation experiments on free draining of a rotating column that nets can dissi-

pate the central air-core vortex
in it when the air-core vortex
touches it due to the interference effects of the nets on the
angular velocity. In a probe
experiment in a cylinder vortex chamber in the present
work, it was shown that a
screen has a similar effect to
nets.

Figure 6.47 illustrates schematically the geometry of the
applied square surface screen.
The with of the screen W_{Screen}
was 240 mm. The l/d ratio
equals 20/4. Suppressing ef-

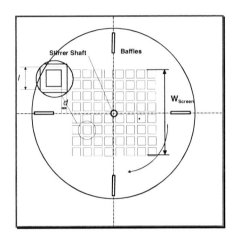

Figure 6.47: Schematic representation of the surface screen

fects depend on the submergence depth of the surface screens. The submergence depth was
set to 70 mm.

The critical rotational speeds were determined in water by visual observations, and it was
found that all critical rotational speeds were increased by about 20%. N_{cloudy} increased from
264.0 up to 313.2 rpm. The surface vortex still existed, but decreased strongly in vortex intensity and depth.

The LDA measurements are summarized in Figure 6.48. Having the similar effect to shaft
baffles, the tangential velocity component decreases significantly (Figure 6.48-a). At the same
time, the RMS value of the tangential velocity is also decreased, indicating the turbulence
weakening effect of the screen. This effect is illustrated more clearly in Figure 6.48-b. The
turbulence kinetic energy drops considerably. However, this decrease results exclusively from
the decrease in the pure turbulence contribution. The dominant frequency and the magnitude
of the MI contribution are almost identical with the case without surface screens. The air entrainment is suppressed, weakening the turbulence level instead of eliminating the surface
vortex. Therefore, as verified in the present work, this possibility of avoiding air entrainment
is also appropriate for other types of impellers.

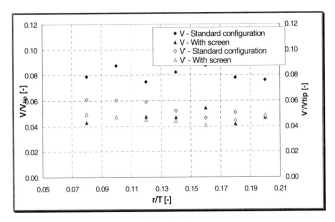

(a): Tangential velocity component

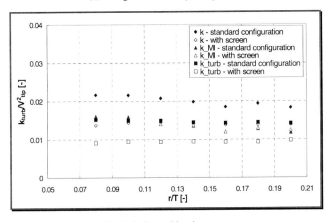

(b): Turbulence kinetic energy

Figure 6.48: Comparison between flow with and without surface screens

7 CONCLUSIONS

The present work was initiated to study the avoidance of surface aeration in stirred vessels. The mechanisms of surface aeration were determined and for the first time associated with the precessing vortex core (PVC) motion and macro instabilities in stirred flows, which have attracted increasing attention in recent decades. The most important results of the present work can be summarized as follows:

Mechanisms of surface aeration and macro instabilities in stirred vessels

- Different types of instabilities occur in rotational flows. An axisymmetric flow tends to lose its stability and become non-axisymmetric, regardless of the axisymmetry of the boundary conditions. PVC is one of the instability consequences in swirl flows which is usually connected with the vortex breakdown. In the present work, the rotating disk flow in a square tank was investigated theoretically, numerically and experimentally. It was shown in the numerical investigation that the flow tends to be unstable if the rotational speed is increased up to a certain level, even without any additional perturbation, and the rotation centre of the flow is put into a precessing motion. This PVC motion was also manifested in the experimental investigations on the rotating disk flow and stirred flows with baffles.

- Air bubble entrainment is induced mainly by the surface vortices and the disturbance of the stirred flow. For radial impellers, the precessing vortex core generates one dominant rotating surface vortex which brings air bubbles at its bottom into the liquid with disturbance from the mean flow. For stirred flow induced by axial impellers, in which the rotational flow is not so pronounced at the surface, one dominant purely radial and non-rotational flow, varying its direction irregularly with time, occurs at the surface and generates vortices behind the baffles downstream. Both mechanisms can be attributed to the macro instability phenomenon.

- Surface aeration in stirred vessels is divided into five sub-processes, in which the surface shape changes to form surface vortices, turbulence level to introduce bubbles into the liquid over surface vortices, forces against surface tension to break up bubbles and sufficient downwards mean flow velocity to drag down the formed bubbles into the stirrer region, play an essential role in the sub-process. A physical model, which takes into account all the mentioned effects, and thus reflects the surface aeration process more correctly, was introduced and verified in the present work. The correlation is expressed as

$$Fr = CWe'^{\alpha}Mo^{\beta}\left(\frac{H-C}{D}\right)^{\gamma}\left(\frac{T}{D}\right)^{\delta}$$

The coefficients have to be determined by experiments for geometries and operating conditions concerned.

Macro instabilities in stirred vessels

- Velocity time series present long-time variations whose time scale and length scale exceed those of turbulence significantly. This type of low variations is induced by the precessing vortex core and penetrates throughout the stirred vessel. The whole flow field induced by impellers involving the periodic blade passing effects is embedded in this type of low variations.

- The low variations can be characterized by spectral analysis. For radial impellers, one dominant frequency exists which corresponds to the rotational frequency of the dominant surface vortex, and hence manifests the inherent association between the PVC and MI in stirred vessels. The constant modified Strouhal number f_{MI}^* represents the independence of Reynolds number and physical properties of stirred media for geometrically similar configurations. For axial impellers, the MI dominant frequencies depend significantly on the measuring time and scatter, indicating a composition of very different frequencies with different dominant levels.

- The magnitude of this type of low variations can be estimated by the decomposition of velocities. It was shown that the magnitude of the MI has a uniform level in the whole vessel, and the contribution of the MI motion to the RMS value formation and the turbulence kinetic energy grows with increasing distance from the turbulence source, the stirrer. This is of great importance for comparisons between experimental and numerical investigations on stirred flows, since the latter have obviously not taken into account the MI contribution to the turbulence kinetic energy, leading to an underestimation of turbulence kinetic energy compared with experimental measurements. The portion of the MI contribution reaches above 80%, indicating that the MI motion is the main mixing mechanism in the uppermost part of stirred tanks.

- The geometry conditions have a significant impact on the dominant frequency. Among them, the impeller bottom clearance has the greatest influence. The dominant frequency leaps between levels of different order. It was shown that for configurations with symmetry around the stirrer the resonant frequency achieves its highest level.

Wall jet flows before baffles

- The wall jet flows before baffles contribute considerably to disturbances on the surface vortices and thus the air bubble entrainment. The wall jet flow can be characterized by the similarity rules to determine the flow regime status. It was shown in the present work that the wall jet flow remains fully turbulent only up to about two-thirds of the tank at a stirrer Reynolds number of 5424 for RT and to half at 7322 for PBT. However, if the MI motion is taken into account, which actually intensifies the wall jet flow varying with time, the wall jet flow at the surface is more turbulent at some instants, disturbing the surface vortices in the form of upwards surges.

Improvement of avoiding surface aeration

- Two types of possibilities can be applied to avoid surface aeration, either to suppress the formation of surface vortices or to weaken the turbulence level surrounding the surface vortices. Two practical possibilities were deduced and verified in the present work: shaft baffles and surface screens. The application of surface screens showed the suppressing effect of the pure turbulence surrounding the surface vortices, which is valid for all types of impellers, whereas shaft baffles have represented both effects for radial impellers.

Recommendations for future work

- The MI motions are, at the same time, responsible for other instantaneous phenomena appearing in stirred flows, e.g., reasonable fluctuations of the macro-mixing time, unstable features of suspension of particles at the bottom, vibrations of the stirrer shaft, and so on. Therefore, more detailed characterization of the MI and its magnitude is of great importance to understand and characterize these unsteady phenomena in stirred flows. This can be carried out first in simple geometries, e.g. in vortex chambers for which Alekseenko et al. [2] were even able theoretically to predict the dominant frequency. This work can also be supported by numerical methods, first of all by LES. As the MI motion can be modelled, these models can be applied to the usual CFD applications in stirred flows to improve the prediction performance of stirred flows.

8 REFERENCES

[1]   Albal, R. S., Shan, Y. T., Schumpe A., 1983, "Mass Transfer in Multiphase Agitated Contactors", *Chem. Eng. J.*, Vol. 27, pp. 61-80.

[2]   Alekseenko, S. V., Kuibin, P.A., Okulov, V. L.& Shtork, S. I., 1998, "Helical Vortices in Swirl Flow", *J. Fluid Mech.*, Vol. 382, pp. 195-243.

[3]   Armenante, P. M., Luo C.-G., Chou, C.-C., Fort, I. & Medek, J., 1997, "Velocity Profiles in a Closed, Unbaffled Vessel: Comparison between Experimental LDV Data and Numerical CFD Predictions", *Chem. Eng. Sci.*, Vol. 52, pp 3483-3492.

[4]   Bakker, A., Oshinowo, L. M. & Marshall, E. M., 2000, "The Use of Large Eddy Simulation to Study Stirred Vessel Hydrodynamics", *Proc. of 10thEurop. Conf. On Mixing*, *Elsevier*, Amsterdam, pp. 247-254.

[5]   Baldyga, J. & Bourne, J., 1999, "Turbulent Mixing and Chemical Reaction", *John Wiley & Sons Ltd.*, Chichester, UK.

[6]   Bartels, C., Breuer, M. & Durst, F., 2000, "Comparison between Direct Numerical Simulation and k-ε Prediction of the Flow in a Vessel Stirred by a Rushton Turbine", *Proc. of 10thEurop. Conf. On Mixing, Elsevier,* Amsterdam, pp. 239-246.

[7]   Bird, R., Stewart, W. & Lightfoot, E. N., 1960, "Transport Phenomena", *Wiley*, New York.

[8]   Bittorf, K. J. & Kresta, S. M., 2000, "Active Volume of Mean Circulation for Stirred Tanks Agitated with Axial Impellers", *Chem. Eng. Sci.*, pp. 1325-1335.

[9]   Bittorf, K. J. & Kresta, S. M., 2000, "Limits of Fully Turbulent Flow in a Stirred Tank", *Proc. of 10thEurop. Conf. On Mixing, Elsevier,* Amsterdam, pp. 17-24.

[10]   Brauer, H. & H. Schmidt-Traub, 1973, "Flüssigkeitsbegasung mit Rührern, Teil 2: Gasgehalt der Sprudelschicht und Belastungsgrenzen des Rührers", *Chem.-Ing.-Tech.*, Vol. 45, pp. 49-52.

[11]   Breuer, M., 2001, "Direkte Numerische Simulation und Large-Eddy Simulation tur-
        bulenter Strömungen auf Hochleistungsrechnern", *Habilitationsschrift*, Technische
        Fakultät, Universität Erlangen-Nürnberg.

[12]   Brodkey, R. S., 1999, "Where Should Mixing Go? A Biased View", *IChemE Symp.*,
        Series No. 146, pp. 1-1.

[13]   Broersen, P.M.T., de Waele, S & Bos, R., 2000, "The Accuracy of Time Series
        Analysis for Laser-Doppler Velocimetry", *10th Int. Symp. on Appl. of Laser Tech. to
        Fluid Mechanics,* Lisbon, July 10-13.

[14]   Bruha, O., Fort, I., Smolka, P., 1994, Flow Transition Phenomenon in an Axially Agi-
        tated System, *Proc. 8th Euro. Conf. Mixing., IChemE Symp. Series No. 136, Cam-
        bridge, UK*, pp. 121-128.

[15]   Bruha, O., Fort, I., Smolka, P., 1995, "Phenomenon of Turbulent Macroinstabilities in
        Agitated Systems", *Collect. Czech. Chem. Commun.,* Vol. 60, pp. 85-94.

[16]   Bruha, O., Fort, I., Smolka, P., Jahode, M., 1996, "Experimental Study of Turbulent
        Macroinstabilities in an Agitated System with Axial High-speed Impeller and with
        Radial Baffles", *Collect. Czech. Chem. Commun.,* Vol. 61, pp. 856-867.

[17]   Calderbank, P. H., 1958, "Physical Rate Processes in Industrial Fermentation - Part I,
        The Interfacial Area in Gas-Liquid Contacting with Mechanical Agitation", *Trans.
        Instn. Chem. Engrs.*, Vol. 36, pp. 443-459.

[18]   Calderbank, P. H., 1959, "Physical Rate Processes in Industrial Fermentation - Part II,
        Mass Transfer Coefficients in Gas-Liquid Contacting with and without Mechanical
        Agitation", *Trans. Instn. Chem. Engrs.*, Vol. 37, pp. 173-185.

[19]   Chanaud, R. C., 1965, "Observation of Oscillatory Motion in Certain Swirling
        Flows", *J. Fluid Mechanics*, Vol. 21, pp.111-127.

[20]   Chandrasekhar, S., 1970, "Hydrodynamic and Hydromagnetic Stability", *Oxford Uni-
        versity Press*, London.

[21]   Chanson, H. & Cummings P, D., 1994, "An Experimental Study on Air Carryunder due to Plunging Liquid Jet – Discussion", *Intl J. of Multiphase Flow*, Vol. 20, pp. 667-770.

[22]   Chanson, H., 1996, "Air Bubble Entrainment in Free-Surface Turbulent Shear Flows", *Academic Press*, London.

[23]   Chapple, D. & Kresta, S. M., 1994, "The Effect of Geometry on the Stability of Flow Patterns in a Stirred Tank", *Chem. Eng. Sci,* Vol. 49, pp. 3651-3660.

[24]   Ciofalo, M., Burcato, A., Grisafi, F. & Torraca, N., 1996, "Turbulent Flow in Closed and Free-surface Unbaffled Tanks stirred by Radial Impellers", *Chem. Eng. Sci.*, Vol. 51, pp. 3557-3573.

[25]   Clark, M., & Vermeulen, T., 1964, "Incipient Vortex Formation in Baffled Agitated Vessels", *AIChE. Journal*, Vol. 10, pp. 420-422.

[26]   Costes J. and Couderc, J. P., 1988, "Study by Laser Doppler Anemometry of the Turbulent Flow Induced by a Rushton Turbine in a Stirred Tank: Influences of the Size of the Units-I. Mean Flow and Turbulence", *Chem. Eng. Sci.*, Vol. 43, pp. 2751-2764.

[27]   Daily, J. W. & Nece, R. E, 1960, "Chamber Dimension Effects on Induced Flow and Frictional Resistance of Enclosed Rotating Discs", *Trans. ASME: J. Basic Engng.*, Vol. 82, pp 217-232.

[28]   Davies, J. T & Rideal, E. K.,1961, "Interfacial Phenomena", *Academic Press*, New York.

[29]   Davies, J. T. & Lozano, F. J., 1979, "Turbulence Characteristics and Mass Transfer at Air-Water Surfaces", *AIChE J.*, Vol. 25, No. 3, pp. 405-415.

[30]   Derksen, J. & Van den Akker, H. E. A., 1999, "Large Eddy Simulations on the Flow Driven by a Rushton Turbine", *AIChE J.*, Vol. 45, pp. 209-221.

[31]   Derksen, J., Doelman, M. S. & Van den Akker, H. E. A., 1999, "Three-Dimensional LDA Measurements in the Impeller Region of a Turbulently Stirred Tank", *Experim. in Fluids*, Vol. 27, pp. 522-532.

[32]   Dierendonck van, L., Fortuin, H. & Venderbos, D., 1971, "The Specific Contact Aera in Gas-Liquid Reactors", *Proc. 4th Europ. Conf. on Chem. Reaction Eng.*, Sept 9-11, Brussels.

[33]   Dijkstra, D. & Heijst, G. J., 1983, "The Flow Between Two Finite Rotating Disks Enclosed by a Cylinder", *J. Fluid Mech.*, Vol. 128, pp 123-154.

[34]   Ditl, P., Rieger, F. & Novak, V., 1997, "Undesirable Air Entrainment in Agitated Baffled Vessels", *Recents progress en Genie des Procedes.*, Vol. 52, pp. 131-136.

[35]   Durst, F. & Schäfer, M., 1996, "A Parallel Block-Structured Multigrid Method for the Prediction of Incompressible Flows", *Int. J. Numer. Meth. Fluids.*, Vol. 22, pp 549-565.

[36]   Durst, F., 2000, "Grundlagen der Strömungsmechanik", Manuscript, Universität Erlangen-Nürnberg.

[37]   Durst, F., Brenn G. & Xu, T. H., 1997, "A Review of the Development and Characteristics of Planar Phase-Doppler Anemometry", *Meas. Sci Technol.*, Vol. 8, pp 1203-1221.

[38]   Durst, F., Melling, A. & Whitelaw, J. H., 1976, "Principles and Practice of Laser-Doppler-Anemometry", *Academic Press*, London.

[39]   Dyster, K. N., Koutsakos, E., Jaworski, Z. and Nienow, A. W., 1993, "An LDA Study of the Radial Discharge Velocities Generated by a Rushton Turbine: Newtonian Fluids, *Re* >= 5", *Trans IChemE,* Vol. 71(A1), pp. 11-23.

[40]   Eggels, J. G. M., 1996, "Direct and Large-Eddy Simulations of Turbulent Fluid Flow Using the Lattice-Boltzmann Scheme", *Int. J. Heat Fluid Flow,* Vol. 17, pp. 307-324.

[41]   Escudier, M. P., 1984, "Observations of the Flow Produced in a Cylindrical Container by a Rotating Endwall", *Exper. In Fluids,* Vol. 2, pp. 189-196.

[42]   Faller, A. J, 1963, "An Experimental Study of the Instability of the Laminar Ekman Boundary Layer", *J. Fluid Mech.*, Vol. 15, pp. 560-576.

[43]   Faller, A. J, 1991, "Instability and Transition of Disturbed Flow over a Rotating Disk", *J. Fluid Mech.*, Vol. 230, pp. 245-269.

[44]   Faller, A. J. & Kaylor, R. E., 1966, "Investigations of Stability and Transition in Rotating Boundary Layers", *Dynamics of Fluids and Plasmas, Academic*, pp. 309-329.

[45]   Ferziger, J. H. and Perić, M., 2002, "Computational Methods for Fluid Dynamics", *Springer*, Berlin.

[46]   Forrester, S. E., Rielly, C.D. & Carpenter K. J., 1997, "Gas-inducing Impeller Design and Performance Characteristics", *Chem. Eng. Sci.*, Vol. 53, pp. 603-615.

[47]   Gad-el-Hak, M., 1988, "Visualization Techniques for Unsteady Flows: An Overview", *J. Fluids Eng.*, Vol. 110, pp. 231-243.

[48]   Gauthier, G., Gondret, P. & Rabaud, M., 1998, "Motions of Anisotropic Particles: Application to Visualization of Three-Dimensional Flow", *Phys. Fluids.*, Vol. 10, pp. 2147-2154.

[49]   Gelfat, A. Y., Bar-Yoseph, P. Z. & Solan, A., 2001, "Three-Dimensional Instability of Axisymmetric Flow in a Rotating Lid-Cylinder Enclosure", *J. Fluid Mech.*, Vol. 438, pp. 363-377.

[50]   Genenger, B. & Wächter P., 1998, "Laboruntersuchungen zur Auslegung eines Rührwerks", *Project Report*, Project Felix Schoeller GmbH &Co. KG, Erlangen.

[51]   Glauert, M. B., 1956, "The Wall Jet", *J. Fluid Mech.*, Vol. 1, pp. 625-643.

[52]   Greaves, M. & Kobbacy, K., 1981, "Surface Aeration in Agitated Vessels", *I. Chem. Symp.* Ser. No. 64, H1.

[53]   Greenspan, H. P., 1968, "The Theory of Rotating Fluids", *Cambridge University Press*, Cambridge, UK.

[54]   Gregory, N., Stuart, J. T., & Walker, W. S., 1955, "On the Stability of Three-Dimensional Boundary Layers with Application to the Flow due to a Rotating Disk", *Phil. Trans. R. Soc. Lond.* Vol. A 248, pp. 155-199.

[55]   Gupta A. K., Lilley D. G. & Syred , N., 1984, "Swirl Flows", *Abacus Press*, Tunbridge Wells, Kent.

[56]   Haam, S., Brodkey, R. S. & Fasano, B., 1992, "Local Heat Transfer in a Mixing Vessel using the Heat Flux Sensors", *Ind. Eng. Chem. Res.,* Vol. 31, pp. 1384-1391.

[57]   Harvey, P. S. & Greaves, M., 1982, "Turbulent Flow in Agitating Vessel, Part I: A Predictive Model", *TRICHE*, Vol. 60, pp. 195-200.

[58]   Hasal, P., Montes, J. L., Boisson, H. C. & Fort, I., 2000, "Macro-Instabilities of Velocity Field in Stirred Vessel: Detection and Analysis", *Chem. Eng. Sci.*, Vol. 55, pp. 391-401.

[59]   Heywood, N., Madhvi, P. & McDonagh, M., 1985, "Design of Ungassed Baffled Mixing Vessels to Minimize Surface Aeration of Low Viscosity Liquids", *5th Eur. Conf. On Mixing*, Würzburg, *10-12 June,1985*, pp. 243-262.

[60]   Hino, M., 1961, "On the Mechanisms of Self-Aerated Flow on Steep Slope Channels", *Proc. 9th IAHR Congress,* pp. 123-132.

[61]   Hinze, J. O., 1975, "Turbulence", *McGraw-Hill,* New York.

[62]   Hockey, R.. M. and Nouri, J. M., 1996, "Turbulent Flow in a Baffled Vessel Stirred by a 60 Degrees Pitched Blade Impeller", *Chem. Eng. Sci.*, Vol. 51, pp. 4405-4421.

[63]   Höfken, M., 1994, "Moderne experimentelle Methoden für die Untersuchung von Strömungen in Rührbehältern und für Rührwerksoptimierungen", PhD Thesis, LSTM-Erlangen, Friedrich-Alexander-Universität Erlangen-Nürnberg.

[64]   Holden, P. J, Wang, M., Mann, R., Dickin, F. J. & Edwards, R. B., 1998, "Imaging Stirred-Vessel Macromixing Using Electrical Resistance Tomography", *AIChE J.*, Vol. 44, pp. 780-190.

[65]   Joshi, J. B., Pandit, A. B. & Sharma, M. M, 1982, "Mechanically Agitated Gas-Liquid Reactors", *Chem. Eng. Sci.*, Vol. 37, pp. 813-844.

[66]   Kamen, A. A., Garnier, A., Andre, G., Archambault, J. & Chavarie, C., 1995, "Determination of Mass Transfer Parameters in Surface Aerated Bioreactors with Bubble Entrainment", *Chem. Eng. J.*, Vol. 59, pp. 187-193.

[67]   Kresta, S. M. & Wood, P. E., 1993, "The Mean Flow Field Produced by 45° Pitched Blade Turbine: Changes in the Circulation Pattern Due to Off Bottom Clearance", *Can. J. Chem. Eng.*, Vol. 71, pp. 42-53.

[68]   Kresta, S. M., Wood, P. E., 1993, "The Flow Field Produced by a Pitched Blade Turbine: Characterization of the Turbulence and Estimation of the Dissipation Rate", *Chem. Eng. Sci.*, Vol. 48, pp. 1761-1774.

[69]   Kudrjawizki, F. & Bauer, M., 1993, "Gastransport durch die Oberfläche gerührter Flüssigkeiten", *Chem.-Ing.-Tech.*, Vol. 59, pp. 187-193.

[70]   Kunte, S., 1996, „Hochaufgelöste Vermessung eines Strömungsfeldes an einem optimierten Rührwerksversuchsstand mittels LDA zur Verifizierung numerischer Simulationen", Diplomarbeit am LSTM-Erlangen, Friedrich-Alexander Universität Erlangen-Nürnberg.

[71]   Landman, M. J., 1990, "On the Generation of Helical Waves in Circular Pipe Flow", *Phys. Fluid A*, Vol. 2, pp. 738-747.

[72]   Leister, H.-J., 1994, "Numerische Simulation dreidimensionaler, zeitabhängiger Strömungen unter dem Einfluß von Auftriebs- und Trägheitskräften", PhD Thesis, LSTM-Erlangen, Friedrich-Alexander-Universität Erlangen-Nürnberg.

[73]   Liepe, F., Sperling, R. & Jembere, S., 1998, "Rührwerke – Theoretische Grundlagen, Auslegung und Bewertung", *Eigenverlag Fachhochschule Köthen*, Köthen.

[74]   Lingwood, R. J., 1995, "Absolute Instability of the Boundary Layer on a Rotating-Disk", *J. Fluid Mech.*, Vol. 299, pp. 17-33.

[75]   Lingwood, R. J., 1997, "Absolute Instability of the Ekman Layer and Related Rotating Flows", *J. Fluid Mech.*, Vol. 331, pp. 405-428.

[76]   Lomb, N. R., 1976, "Least-Squares Frequency Analysis of Unevenly Sampled Data", *Astrophysical J.*, Vol. 39, pp. 447-462.

[77]   Lopez, M., Marques, F. & Sanchez, J., 2001, "Oscillatory Modes in an Enclosed
        Swirling Flow", *J. Fluid Mech.,* Vol. 439, pp. 109-129.

[78]   Lu, W.-M. & Yang, B.-S.,1998, "Effect of Blade Pitch on the Structure of the Trailing
        Vortex around Rushton Turbine Impellers", *Can. J. Chem. Eng.,* Vol. 76, pp. 556-
        562.

[79]   Lugt, Hans J., 1980, "Vortex Flow in Nature and Technique", *Applied Optics,* Vol.
        31, No. 21, pp. 4096-4105.

[80]   Lumley, J., 1978, "Computational Modelling of Turbulent Flows", *Adv. Appl. Mech.,*
        Vol. 26, pp. 123-176.

[81]   Magni, F., Costes, J., Bertrand & Couderc, J. P., 1988, "Study by Laser Doppler
        Anemometry of the Flow by a Rushton Turbine in a Stirred Tank – Influences of the
        Geometry of the Tank Bottom and of the Position of the Turbine", *Proc. of 6th Europ.
        Conf. On Mixing,* Pavia, pp. 7-14.

[82]   Massey, B. S, 1980, "Mechanics of Fluids", *Van Nostrand Reinhold,* New York.

[83]   Matsumura, M., Masunaga, H. & Kobayashi, J., "A Correlation for Flow Rate of Gas
        Entrained from Free Liquid Surface of Aerated Stirred Tank", *J. Ferment. Technol.,*
        Vol. 55, pp. 388-400.

[84]   Mersmann, A., Einenkel, W.-D., Käppel, M., "Auslegung und Maßstabsvergrößerung
        von Rührapparaten", *Chem.-Ing.-Tech.,* Vol. 23, pp. 953 - 964.

[85]   Metzger, A. B. & Taylor, J. S., 1960, "Flow Pattern in Agitated Vessels", *AIChE J.,*
        Vol. 6, pp. 109-114.

[86]   Montes, J. L., Boisson, H. C., Fort, I. & Jahoda, M., 1997, "Velocity Field Macro-
        Instabilities in an Axial Agitated Mixing Vessel", *Chem. Eng. J.,* Vol. 67, pp. 139-
        145.

[87]   Myers, K. J., Ward, R. W. & Bakker, A., 1997, "A Digital Particle Image Veloci-
        metry Investigation of Flow Filed Instabilities of Axial Flow Impellers", *J. Fluids
        Eng.,* Vol. 119, pp. 623-632.

[88]   Nadarajah, S., Balabani, S., Tindal, M. J. & Yianneskis, M., 1998, "The Effect of Swirl on the Annular Flow Past an Axisymmetric Poppet Valve", *Proc Instn Mech Engrs*, Vol. 212 Part C, pp. 473-484.

[89]   Nagata, S., 1975, "Mixing: Principle and Applications", *Wiley*, New York.

[90]   Nouri, J. M. & Whitelaw, H. J., 1990, "Flow Characteristics of Stirred Reactors with Newtonian and non-Newtonian Fluids", *AIChE J.*, Vol. 36, pp. 627-629.

[91]   Padmanabham, G. & Gowda, B. H., 1991, "Mean and Turbulence Characteristics as a Class of Three-Dimensional Wall Jets-Part I: Mean Flow Characteristics", *J. of Fluids Eng.*, Vol. 113, pp. 620-628.

[92]   Park, W., Monticelle, D.A.& White, R. B., 1984, "Reconnection Rates of Magnetic Fields Including the Effects of Viscosity", *Phys. Fluids*, Vol. 27, pp. 137-149.

[93]   Patankar, S., 1980, "Numerical Heat Transfer and Fluid Flow", *Hemisphere*, New York.

[94]   Patwardhan, A., Joshi, J. B., 1998, "Design of Stirred Vessels with Gas Entrained from Free Liquid Surface", *Can. J. Chem. Eng.*, Vol. 76, pp. 339-364.

[95]   Peebles, F. & Garber, H., 1953, "Characteristic Bubble Size in Two-Phase Flows", *Chem. Eng. Progr.*, Vol. 49, pp. 88-92.

[96]   Perić, M., 1988, "Comparison of Finite-Volume Numerical Methods with Staggered and Colocated Grids", *Comput. Fluids*, Vol. 16, pp. 389-403.

[97]   Press, W. H., Flannery, B. P., Teukolsky, S. A. & Vetterling, W. T., 1989, "Numerical Recipes- The Art of Scientific Computing", *Cambridge University Press*, New York. Online: http://www.ulib.org/webRoot/Books/Numerical_Recipes.

[98]   Qiu, H.-H., 1994, "Signalverarbeitung bei der Phasen-Doppler-Anemometrie zur Geschwindigkeits-. Größen- und Konzentrationsmessung in zweiphasigen Strömungen", PhD Thesis, LSTM-Erlangen, Friedrich-Alexander-Universität Erlangen-Nürnberg.

[99]   Qiu, H.-H, Sommerfeld, M. & Durst, F., 1994, "Two Novel Doppler Signal Detection
         Methods for Laser Doppler and Phase Doppler Anemometry", *Meas. Sci. Technol.*,
         Vol. 4, pp. 769-778.

[100]   Ramamurthi, K. & Tharakan, T. J., 1996, "Flow Visualisation Experiments on Free
         Draining of a Rotating Column of Liquid using Nets and Tufts", *Expr. in Fluids*, Vol.
         21, pp. 139-142.

[101]   Rao, A. M. & Brodkey, R. S., 1972, "Continuous Flow Stirred Tank Turbulence Pa-
         rameters in the Impeller Stream", *Chem. Eng. Sci.*, Vol. 27, pp. 137-156.

[102]   Ranade, V. V., Mishra, V. P., Saraph, V. S., Deshpande, G. B. & Joshi, J. B., 1992,
         "Comparison of Axial Flow Impellers Using a Laser Doppler Anemometer", *Ind.
         Eng. Chem. Res.,* Vol. 31, pp. 2370-2379.

[103]   Roussinova, V. T. & Kresta, S. M. 2000, "Analysis of Macro-Instabilities (MI) of the
         Flow Field in Stirred Tank Reactor (STR) Agitated with Different Axial Impellers",
         Proc. of 10thEurop. Conf. On Mixing, Elsevier, Amsterdam, pp. 361-368.

[104]   Roussinova, V. T., Grgic, B., Kresta, S. M. 2000, "Study of Macro-Instabilities in
         Stirred Tanks using a Velocity Decomposition Technique", *Trans IChemE*, Vol. 78,
         pp. 1040-1052.

[105]   Ruck, B., 1987, "Laser-Doppler-Anemometrie", *AT-Fachverlag*, Stuttgart.

[106]   Rushton, J. H. & Sachs, J. P., 1954, "Discharge Flow from Turbine-Type Mixing Im-
         peller", *Chem. Engng.Progr.*, Vol. 50, pp. 597-603.

[107]   Schäfer, M., 1995, „Visualisierung und Charakterisierung turbulenter Rührwerks-
         strömungen mittels moderner Strömungsmeßtechnik", Diplomarbeit am LSTM-
         Erlangen und am Instituto de Hydraúlica der Universidade do Porto, Portugal.

[108]   Schäfer, M., 2001, "Charakterisierung, Auslegung und Verbesserung des Makro- und
         Mikromischens in gerührten Behältern", PhD Thesis, LSTM-Erlangen, Friedrich-
         Alexander-Universität Erlangen-Nürnberg.

[109]   Schäfer, M., Höfken, M. & Durst, F., 1997, "Detailed LDV-Measurements for Visualization of the Flow Field Within a Stirred Tank Reactor Equipped with a Rushton-Turbine", *Trans IChemE*, Vol. 75 (A), pp. 729-736.

[110]   Schäfer, M., Yianneskis, M., Wächter, P. & Durst, F., 1998, "Trailing Vortices Around a 45° Pitched-Blade Impeller", *AIChE Journal*, Vol. 44, pp. 1233-1246.

[111]   Schäfer, M., Yu, J., Genenger, B. & Durst, F., 2000, "Turbulence Generation by different Types of Impellers", *Proc. of 10thEurop. Conf. On Mixing, Elsevier,* Amsterdam, pp. 9-16.

[112]   Schlichting, H., 1979, "Boundary- Layer Theory", *McGraw-Hill*, New York.

[113]   Schouveiler, L., Le Gal, P. & Chauve, M. P., 2001, "Instabilities of the Flow between a Rotating and a Stationary Disk", *J. Fluid Mech.*, Vol. 443, pp. 329-350.

[114]   Serra, A., Campolo, M. & Soldati, A., 2001, "Time-Dependent Finite-Volume Simulation of the Turbulent Flow in a Free-Surface CSTR", *Chem. Eng. Sci.*, Vol. 56, pp. 2715-2720.

[115]   Serre, E. & Bontoux, P., 2001, "Three-Dimensional Swirling Flow with a Precessing Vortex Breakdown in a Rotor-Stator Cylinder", *Phys. Fluids*, Vol. 13, pp 1500-1503.

[116]   Serre, E., Crespo Del Arco, E. & Bontoux, P., 2001, "Annular and Spiral Patterns in Flows between Rotating and Stationary Discs", *J. Fluid Mech.*, Vol. 434, pp 65-100.

[117]   Smith, J. M., 1999, "An Introduction to Mixing Tasks –Challenges and Rewards", *Proc. of the International Seminar on Stirring and Mixing*, 25.-28. October 1999, Lehrstuhl für Strömungsmechanik, Erlangen.

[118]   Stevens, J. L., Lopez, J. M. & Cantwell, B. J., 1999, "Oscillatory Flow States in an Enclosed Cylinder with a Rotating Endwall", *J. Fluid Mech.*, Vol. 389, pp. 101-118.

[119]   Stieglmeier, M. & Tropea, C., 1992, "Mobile Fiber-Optic Laser Doppler Anemometer", *Applied Optics*, Vol. 31, No. 21, pp 4096-4105.

[120]   Stoots, C. M., Calabrese, R. V., 1995, "Mean Velocity Field Relative to a Rushton Turbine Blade", *AIChE Journal*, Vol. 41, pp. 1-11.

[121]   Sverak, S. & Hruby, M., 1981, "Gas Entrainment from the Liquid Surface of Vessels with Mechanical Agitators", *Int. Chem. Eng.*, Vol. 21, pp. 519-526.

[122]   Takase, H., Unno, H., Akehata, T., 1984, "Oxygen Transfer in a Surface Aeration Tank with a Square Cross Section", *Int. Chem. Eng.*, Vol. 24, pp. 128-134.

[123]   Tanaka, M. & Izumi, T., 1987, "Gas Entrainment in Stirred-Tank Reactors", *Chem. Eng. Res. Des,* Vol. 65, pp. 195-198.

[124]   Tanaka, M., Noda, S. & O'shima, E., 1986, "Effect of the Location of a Submerged Impeller on the Enfoldment of Air Bubbles from the Free Surface in a Stirred Vessel", *Int. Chem. Engng,* Vol. 26, pp. 314-318.

[125]   Tatterson, G. B., 1991, "Fluid Mixing and Gas Dispersion in Agitated Tanks", *McGraw-Hill*, New York.

[126]   Tatterson, G. B., 1994, "Scaleup and Design of Industrial Mixing Processes", *McGraw-Hill*, New York.

[127]   Tennekes, H., Lumley, J. L., 1972, „A First Course in Turbulence", *MIT Press*, Cambridge, MA.

[128]   Uhl, V. W, Gray, J. B., 1966, "Mixing – Theory and Practice", *Academic Press*, New York and London.

[129]   User Manual, 2000, "Numeca's Highly Interactive Geometry Modeller and Grid Generator", *Numerical Mechanics Applications International.*

[130]   Van der Molen, K., van Maanen & H. R. E., 1978, "Laser-Doppler-Measurements of the Turbulent Flow in Stirred Vessels to Establish Scaling Rules", *Chem. Eng. Sci.*, Vol. 33, pp. 1161-1168.

[131]   Veljkovic, V. B., Bicok, K., Simonovic, B., 1991, "Mechanism, Onset and Intensity of Surface Aeration in Geometrically Similar, Sparged, Agitated Vessels", *Can. J. Chem. Eng.*, Vol. 69, pp. 916-926.

[132]   Vogel, H. U., 1968, "Experimentelle Ergebnisse über die laminare Strömung in einen zylindrischen Gehäuse mit darin rotierender Scheibe", *Tech. Rep.*, Bericht 6. Max-Plank-Inst.

[133]   Wächter, P., 2000, "Entwicklung von Rührelementen für ein- und mehrphasige Rühraufgaben", PhD Thesis, LSTM-Erlangen, Friedrich-Alexander-Universität Erlangen-Nürnberg.

[134]   Wechsler, K., 2001, "Numerische Berechnungen von Rührwerksströmungen auf Parallelrechnerarchitekturen", PhD Thesis, LSTM-Erlangen, Friedrich-Alexander-Universität Erlangen-Nürnberg.

[135]   Wechsler. K., Breuer, M. & Durst, F., 1999, "Steady and Unsteady Computations of Turbulent Flows Induced by a 4/45° Pitched Blade Impeller", *J. of Fluids Eng., Transactions of the ASME,* Vol. 121, pp. 318-329.

[136]   Wechsler. K., Schäfer, M., Durst, F., 1998, „Advanced Methods for Investigations of Single-Phase Stirred Vessel Flows – Part II: Numerical Methods", Paper No. 238i, *AIChE Annual Meeting,* Miami, USA.

[137]   Wernersson, E., & Trägårdh, Ch., 1998, "Scaling of Turbulence Characteristics in a Turbine-agitated Tank in Relation to Agitator Rate", *Chem. Eng. J,* Vol. 70, pp. 37-45.

[138]   Wichterle, K., Sverak, T., 1996, "Surface Aeration Threshold in Agitated Vessels", *Collect. Czech. Chem. Commun.*, Vol. 61, pp. 681-690.

[139]   Wilcox, D. C., 1993, "Turbulence Modeling for CFD", *California DCW Industries,* California.

[140]   Winardi, S. & Nagase, Y., 1991, "Unstable Phenomenon of Flow in Mixing Vessel with a Marine Propeller", *J. Chem. Eng. Japan,* Vol. 24, pp. 243-249.

[141]   Winardi, S., Nakao, S., Nagase, Y., 1988, "Pattern Recognition in Flow Visualization around a Paddle Impeller", *J. Chem. Eng . Japan*, Vol. 21, pp. 243-249.

[142]   Wu, H., Patterson, G.K., 1989, "Distribution of Turbulence Energy Dissipation Rates in a Rushton turbine Stirred Mixer", *Chem. Eng. Sci.*, Vol. 44, pp. 2207-2221.

[143]   Yazdabadi, P. A., Griffiths, A. J. & Syred N., 1994, "Characterization of the PVC
         Phenomena in the Exhaust of a Cyclone Dust Separator", *Experiments in Fluids*, Vol.
         17, pp. 84-95.

[144]   Yianneskis, M., 2000, „Trailing Vortex, Mean Flow and Turbulence Modification
         Through Impeller Blade Design in Stirred Reactors", *Proc. of 10ᵗʰEurop. Conf. On
         Mixing, Elsevier,* Amsterdam, pp. 1-8.

[145]   Yianneskis, M., Popiolek Z. & Whitelaw J.H., 1987, "An Experimental Study of the
         Steady and Unsteady Flow Characteristics of Stirred Reactors", *J. Fluid Mech.*, Vol.
         175, pp. 537-555.

[146]   Yianneskis, M. & Whitelaw J.H., 1993, "On the Structure of the Trailing Vortices
         around Rushton Turbine Blades", *Trans IChemE*, Vol. 71 (A), pp. 543-550.

[147]   Yu, Z. & Rasmuson, A., 2000, "Characterization and Rotation Symmetry of the Im-
         peller Region in Baffled Agitated Suspensions", *Proc. of 10ᵗʰEurop. Conf. On Mixing,
         Elsevier,* Amsterdam, pp. 423-430.

[148]   Zandbergen, P. J. & Dijkstra, D., 1987, "Von Karman Swirling Flows", *Ann. Rev.
         Fluid Mech.*, Vol. 19, pp. 465-461.

[149]   Zehner, P. & Kraume, M., 1999, "Surface Aeration in Stirred Vessels", *Two-Phase
         Flow Modelling and Experimentation 1999*, Edizioni ETS, Pisa, pp. 1289-1296.

[150]   Zlokarnik, M., 1979, "Scale-up of Surface Aerator for Waste Water Treatment", *Adv.
         Biochem. Eng.*, Vol 11., pp. 158-180.

[151]   Zlokarnik, M., 1991, "Dimensional Analysis and Scale-up in Chemical Engineering",
         Springer-Verlag, Berlin, Heidelberg.

[152]   Zlokarnik, M., 1999, "Rührtechnik - Theorie und Praxis", *Springer-Verlag*, Berlin
         Heidelberg.

APPENDICES

Appendix A: Runge-Kutta Method with Newton-Raphson searching method

<u>Introduction of equations</u>

The system of ordinary equations (ODE) being solved in Section 4.2.2 is listed as following:

$$2F + H' = 0, \tag{A.1}$$

$$F^2 + F'H - G^2 - F'' = 0, \tag{A.2}$$

$$2FG + HG' - G'' = 0. \tag{A.3}$$

With the boundary conditions as

$$\begin{aligned} \zeta = 0: & \quad F = 0, \quad G = 1, \quad H = 0, \\ \zeta = \infty: & \quad F = 0, \quad G = 0. \end{aligned} \tag{A.4}$$

In Equation (A.2) and (A.3) second-order derivates of F and G are involved in the ordinary differential equation system. An equation with higher order derivatives which can be expressed as $y^{(n)} = f(x, y, y', ..., y^{(n-1)})$ can be rewritten as a system of first-order equations by making the substitutions as

$$y_1 = y, \ y_2 = y', ..., y_n = y^{(n-1)}. \tag{A.5}$$

Then

$$\begin{aligned} y' &= y_2, \\ y_2' &= y_3, \\ &\quad . \\ &\quad . \\ y_n' &= f(x, y_1, y_2, ..., y_n). \end{aligned} \tag{A.6}$$

is a system of n first-order ODEs.

Correspondingly, Equation (A.1)-(A.3) can be rearranged as

$$F' = F_1, \tag{A.7}$$

$$F_1' = F^2 - G^2 + F_1 H, \tag{A.8}$$

$$G' = G_1, \tag{A.9}$$

$$G_1' = 2FG + HG_1, \tag{A.10}$$

$$H' = -2F \tag{A.11}$$

Fourth-order Runge-Kutta method

The formula for the Euler method is

$$y_{n+1} = y_n + hf(x_n, y_n), \tag{A.12}$$

which advances a solution from x_n to $x_{n+1} \equiv x_n + h$. The Euler method is not recommended for practical use, since the method is neither accurate ($O(h^2)$) nor stable. Runge-Kutta method takes a "trial" step to the midpoint of the interval. In equations,

$$
\begin{aligned}
k_1 &= hf(x_n, y_n), \\
k_2 &= hf(x_n + \frac{1}{2}h, y_n + \frac{1}{2}k_1),, \\
y_{n+1} &= y_n + k_2 + O(h^3).
\end{aligned}
\tag{A.13}
$$

We get the equations of a second-order Runge-Kutta method. Note that, a method is conventionally called nth order if its error term is $O(h^{n+1})$. The most often used is the classic fourth-order Runge-Kutta formula which has a certain sleekness of organization about it:

$$
\begin{aligned}
k_1 &= hf(x_n, y_n), \\
k_2 &= hf(x_n + \frac{1}{2}h, y_n + \frac{1}{2}k_1), \\
k_3 &= hf(x_n + \frac{1}{2}h, y_n + \frac{1}{2}k_2), \\
k_4 &= hf(x_n + h, y_n + k_3), \\
y_{n+1} &= y_n + \frac{k_1}{6} + \frac{k_2}{3} + \frac{k_3}{3} + \frac{k_4}{6} + O(h^5).
\end{aligned}
\tag{A.14}
$$

The fourth-order Runge-Kutta method requires four evaluations of the right hand side per step h.

Newton-Raphson searching method for two-point boundary value problems

The boundary conditions for Equation (A.7)-(A.11) are given by Equation (A.4). Two initial values are missing, namely the $F'(0)$ and $G'(0)$. As a compensation, two boundary conditions at the endpoints are given for the equation system. The shooting method is necessary to calculate the initial value of F' and G'.

Assume that at the starting point x_1 there are N starting values y_i to be specified, but subject to n_1 conditions. Therefore there are $n_2 = N - n_1$ freely specified starting values. Imagine that there freely specifiable values are the components of a vector $\mathbf{V}$ that lives in a vector space of dimension n_2. Then, a function that generates a complete set of N starting values $\mathbf{y}$ can be written, satisfying the boundary conditions at x_1, form an arbitrary vector value of $\mathbf{V}$ in which there are no restrictions on the n_2 component values:

$$y_i(x_1) = y_i(x_1; V_1, ..., V_{n_2}) \qquad i = 1, ..., N, \tag{A.15}$$

this function is also normally called "load".

Given a particular $\mathbf{V}$, a particular $\mathbf{y}(x_1)$ is thus generated. It can be turned into a $\mathbf{y}(x_2)$ by integrating the ODEs to x_2 as an initial value problem. Now, we can define a discrepancy vector $\mathbf{F}$, also of dimension n_2, whose components measure how far we are form satisfying the n_2 boundary condition at x_2:

$$F_k = B_{2k}(x_2, \mathbf{y}) \qquad k = 1, ..., n_2, \tag{A.16}$$

The so-called "score" procedure try to find a vector value of $\mathbf{V}$ that zeros the vector of $\mathbf{F}$ recalling the heart of the Newton-Raphson method:

$$\mathbf{J} \cdot \delta\mathbf{V} = -\mathbf{F}, \tag{A.17}$$

provided that the boundary condition at x_2 are zero, and then adding the correction back,

$$\mathbf{V}^{new} = \mathbf{V}^{old} + \delta\mathbf{V}. \tag{A.18}$$

The Jacobian matrix $\mathbf{J}$ has components given by

$$J_{ij} = \frac{\partial F_i}{\partial V_j}. \tag{A.19}$$

It is not feasible to compute these partial derivatives analytically. Rather, each requires a separate integration of the N ODEs, followed by the evaluation of

$$\frac{\partial F_i}{\partial V_j} \approx \frac{F_i(V_1, ..., V_j + \Delta V_j, ...) - F(V_1, ..., V_j, ...)}{\Delta V_j}. \tag{A.20}$$

Results

The ODE solver used in the present work is an adaptive stepsize controlled, fourth-order Runge-Kutta integrator. So the whole procedure to solve the system (A.1)-(A.4) is:

(i)   Give a initial value to F' and G', namely V_1 and V_2,

(ii)   Solve (A.1)-(A.4) by fourth-order Runge-Kutta method. If $\left|F'(\infty)\right|$, $\left|G'(\infty)\right| \leq \varepsilon_{res}$, end the computation. ($\varepsilon_{res}$ is an user predescribing error), else go to (iii).

(iii)   Let $\mathbf{V} = \mathbf{V} + \delta\mathbf{V}$, solve Equation (A.1)-(A.4) by fourth-order Runge-Kutta method.

(iv)   Compute $\mathbf{J}$ by Formula (A.19)-(A.20).

(v)   Compute $\delta\mathbf{V}$ by solving system (A.17).

(vi)   Let $\mathbf{V} = \mathbf{V} + \delta\mathbf{V}$, go to (ii).

The guessed initial values are both set to 0. The result is listed in Table A.1 and also depicted in Figure 4.3.

$\zeta = z\sqrt{\dfrac{\omega}{v}}$	F'	$-G'$
0	0.510	0.6159

$\zeta = z\sqrt{\dfrac{\omega}{v}}$	F	G	H
0.0000	0.0000	1.0000	0.0000
0.0001	0.0001	0.9999	0.0000
0.0002	0.0001	0.9999	0.0000
0.0003	0.0002	0.9998	0.0000
0.0004	0.0002	0.9998	0.0000
0.0009	0.0005	0.9995	0.0000
0.0014	0.0007	0.9992	0.0000
0.0019	0.0010	0.9988	0.0000
0.0024	0.0012	0.9985	0.0000
0.0048	0.0025	0.9970	0.0000
0.0073	0.0037	0.9955	0.0000
0.0098	0.0049	0.9940	0.0000

0.0245	0.0122	0.9849	-0.0003
0.0368	0.0181	0.9773	-0.0007
0.0492	0.0239	0.9697	-0.0012
0.0615	0.0295	0.9622	-0.0019
0.1198	0.0543	0.9265	-0.0068
0.1781	0.0761	0.8912	-0.0144
0.2364	0.0952	0.8563	-0.0244
0.2947	0.1119	0.8221	-0.0365
0.4006	0.1364	0.7617	-0.0629
0.5066	0.1545	0.7041	-0.0938
0.6126	0.1671	0.6494	-0.1280
0.7186	0.1751	0.5979	-0.1644
0.8548	0.1800	0.5363	-0.2129
0.9911	0.1801	0.4801	-0.2620
1.1274	0.1765	0.4289	-0.3107
1.2637	0.1703	0.3827	-0.3580
1.4238	0.1607	0.3341	-0.4110
1.5838	0.1496	0.2913	-0.4607
1.7439	0.1377	0.2537	-0.5067
1.9040	0.1256	0.2208	-0.5489
2.0679	0.1134	0.1913	-0.5880
2.2318	0.1018	0.1657	-0.6233
2.3958	0.0908	0.1434	-0.6548
2.5597	0.0806	0.1240	-0.6829
2.7439	0.0702	0.1053	-0.7107
2.9281	0.0609	0.0893	-0.7348
3.1123	0.0526	0.0757	-0.7556
3.2965	0.0453	0.0642	-0.7737
3.5171	0.0377	0.0526	-0.7919
3.7378	0.0313	0.0430	-0.8071
3.9584	0.0259	0.0351	-0.8197
4.1790	0.0214	0.0286	-0.8301

Table A.1: Values of the functions calculated by the Runge-Kutta method

Appendix B: Solutions of ODEs with FV Method Applying TDMA

Discretization of the differential equations

With the help of similarity solutions, the system of Navier-Stokes equations is simplified into a system of ordinary differential equations which are listed as following:

$$2F + H' = 0, \tag{B.1}$$

$$F^2 + F'H - G^2 - F'' = 0, \tag{B.2}$$

$$2FG + HG' - G'' = 0. \tag{B.3}$$

This equation system can be considered as steady one-dimensional ordinary differential equations. To solve the equation system with FV method, the calculation domain is divided into a number of nonoverlapping control volumes such that there is one control volume surrounding each grid points as depicted in Figure B.1. We focus attention on the grip point P, which has the grid points E and W as its neighbours (E denotes the east side, i.e., the positive x direction, while W stands for west or the negative x direction.). The dashed lines show the faces of the control volume, and the letter e and w denote these faces. For the one-dimensional problem under consideration, the thickness in the other two spatial directions is assumed to be unit. Thus the volume of the control volume shown is $\Delta x \times 1 \times 1$.

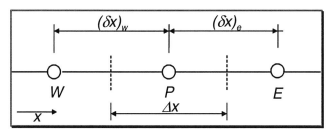

Figure B.1: Grid-point cluster for the one-dimensional problem

Substituting Equation (B.1) into Equation (B.2), and integrating Equation (B.2) and (B.3), it gives:

$$\int_w^e F^2 dx + FH\big|_w^e + 2\int_w^e F^2 dx - \int_w^e G^2 dx - F'\big|_w^e = 0 \tag{B.4}$$

$$\int_w^e 2FG dx + HG\big|_w^e + \int_w^e 2FG dx - G'\big|_w^e = 0 \tag{B.5}$$

Furthermore, the equations can be rewritten as

$$3\int_w^e F^2 dx + FH\Big|_w^e - \int_w^e G^2 dx = \left(\frac{dF}{dx}\right)_e - \left(\frac{dF}{dx}\right)_w \tag{B.6}$$

$$4\int_w^e FG dx + HG\Big|_w^e = \left(\frac{dG}{dx}\right)_e - \left(\frac{dG}{dx}\right)_w \tag{B.7}$$

Ultimately, the equations can be discretized as following:

$$\frac{F_E - F_P}{\Delta x} - \frac{F_P - F_W}{\Delta x} + G^2 \Delta x - (FH)_e + (FH)_w - 3F^2 \Delta x = 0 \tag{B.8}$$

$$\frac{G_E - G_P}{\Delta x} - \frac{G_P - G_W}{\Delta x} - (HG)_e + (HG)_w - 4FG \Delta x = 0 \tag{B.9}$$

Solution of the Linear Algebraic Equations

The solution of the discretization equations for the one-dimensional situation can be obtained by the standard-Gaussian-elimination method. Because of the particularly simple form of the equations, the elimination process turns into a delightfully convenient algorithm. This is often called the TDMA (*Tri*Diagonal-*M*atrix *A*lgorithm). The designation TDMA refers to the fact that when the matrix of the coefficients of these equations is written, all the nonzero coefficients align themselves along three diagonals of the matrix.

For convenience in presenting the algorithm, it is necessary to use somewhat different nomenclature. According to the grid points in Figure B.1, the discretization equations can be written in a numbered form as

$$a_i T_i = b_i T_{i+1} + c_i T_{i-1} + d_i, \tag{B.10}$$

for $i = 1, 2, 3, \ldots, N$. The term T_i to be calculated is related to the neighbouring points T_{i+1} and T_{i-1} in such a way. Taking into account the boundary conditions (i.e. T_1 or/and T_N are given, for example if T_1 is given, we have $a_1 = 1$, $b_1 = 1$, $c_1 = 0$, and $d_1 =$ given value of T_1), T_0 and T_{N+1} will not have any meaningful role to play. In the forward-substitution process, there is a relation

$$T_i = P_i T_{i+1} + Q_i, \tag{B.11}$$

after we have obtained

$$T_{i-1} = P_{i-1}T_i + Q_{i-1}. \tag{B.12}$$

Substitution of Equation (B.12) into Equation (B.10) leads to

$$a_iT_i = b_iT_{i+1} + c_i(P_{i-1}T_i + Q_{i-1}) + d_i, \tag{B.13}$$

which can be rearranged to look like Equation (B.11). In other words, the coefficients P_i and Q_i then stand for

$$P_i = \frac{b_i}{a_i - c_iP_{i-1}}, \tag{B.14}$$

$$Q_i = \frac{d_i + c_i}{a_i - c_iP_{i-1}}. \tag{B.15}$$

These are recurrence relations, since they give P_i and Q_i in terms of P_{i-1} and Q_{i-1}. To start the recurrence process, we note that Equation (B.10) for $i = 1$ is almost of the form of Equation (B.11). Thus, the values of P_1 and Q_1 are given by

$$P_1 = \frac{b_1}{a_1} \quad \text{and} \quad Q_1 = \frac{d_1}{a_1}. \tag{B.16}$$

At the other end of the P_i, and Q_i sequence, we note that $b_N = 0$. This leads to $P_N = 0$, and hence from Equation (B.11) we obtain

$$T_N = Q_N. \tag{B.17}$$

Afterwards, we are in a position to start the back substitution via Equation (B.11).

The tridiagonal-matrix algorithm is a very powerful and convenient equation solver whenever the algebraic equations can be represented in the form of Equation (B.10). Unlike general matrix methods, the TDMA requires computer storage and computer time proportional only to N, rather than to N^2 and N^3.

Results

The discretization equations can be rewritten in a similar form as Equation (B.10)

$$\frac{2}{\Delta x} \cdot F_P = \frac{1}{\Delta x} \cdot F_E + \frac{1}{\Delta x} \cdot F_W + G^2 \Delta x - (FH)_e + (FH)_w - 3F^2 \Delta x \qquad (B.18)$$

$$\frac{2}{\Delta x} \cdot G_P = \frac{1}{\Delta x} \cdot G_E + \frac{1}{\Delta x} \cdot G_W - (HG)_e + (HG)_w - 4FG\Delta x \qquad (B.19)$$

For Equation (B.18), the coefficients are given as

$$a = \frac{2}{\Delta x}, \ b = \frac{1}{\Delta x}, \ c = \frac{1}{\Delta x} \ and \ d = G^2 \Delta x - (FH)_e + (FH)_w - 3F^2 \Delta x \qquad (B.20)$$

In a similar way, the coefficients of Equation (B.19) can be expressed as

$$a = \frac{2}{\Delta x}, \ b = \frac{1}{\Delta x}, \ c = \frac{1}{\Delta x} \ and \ d = -(HG)_e + (HG)_w - 4FG\Delta x \qquad (B.21)$$

Substituting the coefficients into the equations, and assuming the initial values at all points with a line, the iteration procedure converges to the ultimate results. The grid resolution was set as 1/30, and the residual was chosen as 0.0001. The results are listed in Table B.1 and Figure B.2.

$\zeta = z\sqrt{\dfrac{\omega}{\nu}}$	F'	$-G'$	$-H$
0	0.491	0.614	0

Table B.1: Values of the functions calculated by the TDMA method

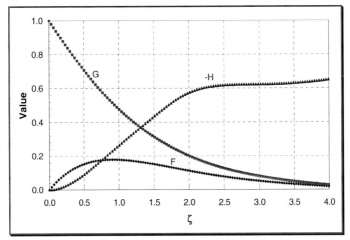

Figure B.2: Values of the functions calculated by the TDMA method

In a similar way, the Boedewadt layer which is generated when the fluid at a large distance above a stationary wall rotates at a constant angular velocity, can be solved. The flow is shown in perspective by Figure B.3. The equations are listed as:

$$F^2 - G^2 + HF' - F'' + 1 = 0,$$ (B.22)

$$2GF + HG' - G'' = 0,$$ (B.23)

$$2F + H' = 0.$$ (B.24)

With the boundary conditions

$$\zeta = 0: \quad F = 0; \quad G = 0; \quad H = 0;$$
$$\zeta = \infty: \quad F = 0; \quad G = 1.$$ (B.25)

and the solution is shown in Table B.2 and Figure B.4.

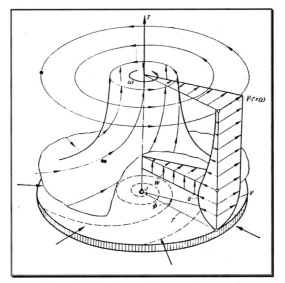

Figure B.3: Rotating of flow near the ground

$\zeta = z\sqrt{\dfrac{\omega}{\nu}}$	F'	G'	H
0	-0.891	0.772	0

Table B.2: Values of the functions calculated by the TDMA method for Boedewadt layer

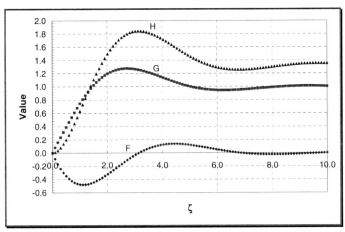

Figure B.4: Values of the functions calculated by the TDMA method for Boedewadt layer

Appendix C: Lomb Spectral Algorithm for Unevenly Sampled LDA Data

If a continuous signal $x(t)$ is discretized, it can be written as $x(t) = x(t_i)$, where $i = 1,...,N$. For evenly sampled data, it is valid that $x(t_i) = x(n \cdot \Delta)$, where $n = ...,-3, -2, -1, 0, 1, 2, 3,...$, where Δ is the sampling interval, whose reciprocal is the sampling rate. The Nyquist critical frequency defined as following is of great importance:

$$f_c \equiv \frac{1}{2\Delta}. \tag{C.1}$$

Complete information about all spectral components is contained in a signal $x(t_i)$ up to the Nyquist frequency, and information about any signal components at frequencies larger than this is scrambled or aliased.

Now define that $b_i(t)$ is an orthogonal basis set which defines the transform, the coefficients $c(i)$ which represent the $x(t)$ in the transformation domain read:

$$c(i) = \int_{-\infty}^{+\infty} x(t) b_i(t) dt. \tag{C.2}$$

Calculated coefficients are those which minimize the squared error defined as

$$\varepsilon = \int_{-\infty}^{+\infty} \left(x(t) - c(i) b_i(t) \right)^2 dt. \tag{C.3}$$

In case of the evenly sampled data in the Fourier domain Equation (C.3) is the well known discrete time Fourier transformation (DTFT), and the fast algorithm used to compute is the Fast Fourier Transformation (FFT). For the signal which can be only acquired at unevenly spaced time instants, Lomb proposed to estimate the Fourier spectra by adjusting the model given as

$$x(t_n) + \varepsilon_n = a\cos(2\pi f_i t_n) + b\sin(2\pi f_i t_n). \tag{C.4}$$

Here the mean square error ε_n is minimized by the proper selection of parameters a and b. Equation (C.4) can be easily proven to be a particular case for real signals from the more general formulation given as

$$x(t_n) + \varepsilon_n = c(i) e^{j 2\pi f_i t_n}, \tag{C.5}$$

where the $x(t)$ and $c(i)$ can be complex values. For any transformation including the Fourier Transformation, the expression will be reformed as

$$x(t_n) + \varepsilon_n = c(i) b_i(t_n). \tag{C.6}$$

Minimization of ε_n variance (root mean square error) leads to minimization of

$$\sum_{n=1}^{N} \left| x(t) - c(i) b_i(t_n) \right|^2 . \tag{C.7}$$

The resulting value for $c(i)$ should be

$$c(i) = \frac{1}{k} \sum_{n=1}^{N} x(t_n) b_i^*(t_n) , \tag{C.8}$$

and k is defined by $k = \sum_{n=1}^{N} \left| b_i(t_n) \right|^2 .$

The Lomb normalized periodogram, namely spectral power as a function of angular frequency $\omega = 2\pi f > 0$, can be defined by

$$P_N(\omega) = \frac{1}{2\sigma^2} \left\{ \frac{\left[\sum_j (x(t_j) - \overline{x}) \cos(\omega(t_j - \tau)) \right]^2}{\sum_j \cos^2(\omega(t_j - \tau))} + \frac{\left[\sum_j (x(t_j) - \overline{x}) \sin(\omega(t_j - \tau)) \right]^2}{\sum_j \sin^2(\omega(t_j - \tau))} \right\} . \tag{C.9}$$

The mean and the variance of the signal $x(t_n)$ and the constant τ are given by

$$\overline{x} = \frac{1}{N} \sum_{i=1}^{N} x(t_i), \quad \sigma^2 = \frac{1}{N-1} \sum_{i=1}^{N} (x(t_i) - \overline{x})^2, \quad \tan(2\omega\tau) = \frac{\sum_j \sin 2\omega t_j}{\sum_j \cos 2\omega t_j} . \tag{C.10}$$

The constant τ is a kind of offset that makes $P_N(\omega)$ completely independent of shifting all the t_i's by any constant. It makes Equation (C.9) identical to the equation obtained if one estimated the harmonic content of a data set, at given frequency ω, by linear least squares fitting to the model

$$h(t) = A \cos \omega t + B \sin \omega t . \tag{C.11}$$

From above it can be seen that the Lomb algorithm weights the data on a "per point" basis instead of on a "per time interval" basis. Assume that $P_N(\omega)$ has an exponential probability distribution with unit mean, namely the probability that $P_N(\omega)$ will be between some positive z and $z+dz$ is $exp(-z)dz$. It readily follows that, if some M independent frequencies are scanned, the probability that none give values larger than z is $(1 - e^{-z})^M$. So

$$P(> z) \equiv 1 - (1 - e^{-z})^M . \tag{C.12}$$

is the false-alarm probability of the null hypothesis, the significance level of any peak in $P_N(\omega)$. In general M depends on the number of frequencies sampled, the number of data points N, and their detailed spacing. It turns out that M is very nearly equal to N when the data points are approximately equally spaced, and when the sampled frequencies oversample the frequency range from 0 to the Nyquist frequency f_c.

The program **PERIOD** in [97] implements the Lomb method described above. For more details the reader can refer to *Press et al.* [97]. The Lombe algorithm is slow, it requires $N_\omega N$ operations in order to calculate N_ω frequencies from N data points. Thus the most evaluations were carried out on an available workstation.

Gaseintrag und Makroinstabilität
in gerührten Behältern

Der Technischen Fakultät der
Universität Erlangen-Nürnberg

zur Erlangung des Grades

DOKTOR – INGENIEUR

vorgelegt von

Jian Yu

Erlangen - 2003

Printed with the support of
German Academic Exchange Service

Als Dissertation genehmigt von
der Technischen Fakultät der
Universität Erlangen-Nürnberg

Tag der Einreichung: 24. Juni 2002
Tag der Promotion: 17. Januar 2003
Dekan: Prof. Dr. rer. nat. A. Winnacker
Brichterstatter: Prof. Dr. Dr. h.c. F. Durst
                              Prof. Dr.-Ing. K.-E. Wirth

INHALTSVERZEICHNIS

VORWORT VII

KURZFASSUNG VIII

INHALTSVERZEICHNIS IX

SYMBOLVERZEICHNIS XIII

1   EINLEITUNG 1

  1.1 ALLGEMEINE BETRACHTUNGEN 1
  1.2 KURZER LITERATURSTAND 3
    1.2.1 Der Mechanismus der Oberflächenbegasung 4
    1.2.2 Korrelationsansätze des Einbruchs der Oberflächebegasung in gerührten Behältern 6
    1.2.3 Makroinstabilitäten in gerührten Behältern 9
  1.3 INHALT DER DISSERTATION 11

2   THEORETISCHE GRUNDLAGEN 13

  2.1 ROTIERENDE STRÖMUNGEN 13
    2.1.1 Grundgleichungen 13
    2.1.2 Dimensionsanalyse und physikalische Ähnlichkeit 15
    2.1.3 Klassifizierung der Wirbel 17
    2.1.4 Sekundäre Strömung bei rotierenden Körpern 18
    2.1.5 Wirbelbehaftete Strömung und die Wirbelablösung 19
    2.1.6 Rotierende Strömung mit freier Oberfläche 20
    2.1.7 Instabilitäten in rotierenden Strömungen 21
      2.1.7.1   Hydrodynamische Instabilitäten 21
      2.1.7.2   Klassifizierung der Instabilitäten in rotierenden Scheibenströmungen 22
    2.1.8 PVC in wirbelbehafteten Strömungen 28
  2.2 TURBULENZ UND RÜHREN 30
    2.2.1 Turbulente Strömungen und Turbulenzmodellierung 30
    2.2.2 Strömungsfelder in gerührten Behältern 32
    2.2.3 Charakterisierung der gerührten Strömungen 33
    2.2.4 Turbulenzverteilung in gerührten Behältern 35
  2.3 GASEINTRAG ÜBER FREIE OBERFLÄCHEN 36
    2.3.1 Einleitung 36
    2.3.2 Mechanismus des Gaseintrags 36

3   EINGESETZTE EXPERIMENTELLE UND NUMERISCHE VERFAHREN 39

  3.1 STRÖMUNGSSICHTBARMACHUNG MIT LASERLICHTSCHNITT UND VIDEOBEARBEITUNG 39
  3.2 LASER-DOPPLER-ANEMOMETRIE 40
    3.2.1 Grundlagen der LDA 40
    3.2.2 Eingesetztes LDA-System 41
      3.2.2.1   Experimenteller Aufbau 41

3.2.2.2   Messstrecken 41

3.2.2.3   LDA-Messsystem und Messdatenerfassung 44

3.2.2.4   Festlegung der notwendigen Anzahl von Messwerten und Genauigkeit der Messung 47

3.3   NUMERISCHE METHODEN UND STRÖMUNGSBERECHUNG 48

3.3.1 Allgemeine Einführung 48

3.3.2 Finite-Volumen Methode 49

3.3.2.1   Räumliche Diskretisierung 49

3.3.2.2   Diskretisierung in der Zeit 50

3.3.3 Gittererzeugung 51

4   EINFÜHRUNGSUNTERSUCHUNGEN ZU ROTIERENDEN SCHEIBENSTRÖMUNGEN 53

4.1   GEOMETRIE 53

4.2   ANALYTISCHE UNTERSUCHUNGEN ZUR ROTIERENDEN SCHEIBENSTRÖMUNG 54

4.2.1 Einführung 54

4.2.2 Analytische Lösungen 54

4.3   NUMERISCHE UNTERSUCHUNGEN DER ROTIERENDEN SCHEIBENSTRÖMUNG 57

4.3.1 Einführung 57

4.3.2 Gitterblockstruktur 58

4.3.3 Ergebnisse und Diskussionen 58

4.3.3.1   Stationäre Strömungsberechnungen 58

4.3.3.2   Instationäre Strömungsberechnungen 64

4.4   EXPERIMENTELLE UNTERSUCHUNGEN ZUR ROTIERENDEN SCHEIBENSTRÖMUNG 67

4.4.1 Visualisierung der rotierenden Scheibenströmung in einem Behälter mit quadratischer Grundfläche .. 67

4.4.2 Stationäre Strömungsuntersuchungen in gerührten Behältern 69

4.4.2.1   Einführung und experimenteller Aufbau 69

4.4.2.2   Ergebnisse und Diskussion 69

5   ERGEBNISSE DER VISUELLEN BEOBACHTUNGEN DER OBERFLÄCHENBEGASUNG 73

5.1   ALLGEMEINE MECHANISMEN DES GASEINTRAGS IN GERÜHRTEN BEHÄLTERN 73

5.1.1 Unbewehrte Behälter 73

5.1.2 Bewehrte Standardbehälter 75

5.1.2.1   Experimenteller Aufbau 75

5.1.2.2   Allgemeine Mechanismen des Gaseintrags 75

5.1.2.3   Andere Mechanismen des Gaseintrags 78

5.1.3 Lokalisierung der Oberflächenwirbel in bewehrten Behältern 79

5.1.3.1   Radialfördernde Rührelemente 79

5.1.3.2   Axialfördernde Rührelemente 81

5.1.4 Visualisierung der Struktur von Oberflächenwirbeln 82

5.1.4.1   Einführung und Geometrie 82

5.1.4.2   Ergebnisse und Diskussionen 83

5.2   KRITISCHE DREHZAHLEN DES EINBRUCHS DER OBERFLÄCHENBEGASUNG 85

5.2.1 Bestimmung der unterschiedlichen kritischen Drehzahlen 85

5.2.2 Vergleich mit vorhandenen Korrelationen 86

5.2.3 Theoretische Analyse der Korrelationen 88

6   UNTERSUCHUNGEN DER OBERFLÄCHENBEGASUNG UND MAKROINSTABILITÄT 93

6.1   UNTERSUCHUNGEN ZUM STRÖMUNGSFELD 93

6.1.1 Strömungsparameter *93*

6.1.2 Unbewehrte und Nicht-Voll-Bewehrte gerührte Strömungen *95*

6.1.3 Voll-Bewehrte gerührte Strömungen *96*

    6.1.3.1   Sechs-Blatt-Scheibenrührer 96

    6.1.3.2   Schrägblattrührer 102

6.2 UNTERSUCHUNGEN ZUR MAKROINSTABILITÄT 104

6.2.1 Zeitreihenanalyse der Geschwindigkeiten *105*

6.2.2 Spektrale Analyse von Makroinstabilitäten *107*

    6.2.2.1   Lomb Periodogramalgorithm 107

    6.2.2.2   Räumliche Verteilung der dominanten MI-Frequenz 109

    6.2.2.3   Linearität der dominanten MI-Frequenz 111

    6.2.2.4   Einfluss der Messrate auf die Bestimmung der MI 115

    6.2.2.5   Einfluss der Messzeit auf die Bestimmung der MI 115

6.2.3 Intensität der Makroinstabilitäten *119*

    6.2.3.1   Einführung 119

    6.2.3.2   Dekompositionsmethode 120

    6.2.3.3   Verfahren des bewegten Mittelungsfensters 121

    6.2.3.4   Berechung der Intensität 123

    6.2.3.5   Räumliche Verteilung der Intensität niedriger Variationen 125

    6.2.3.6   Abhängigkeit der MI-Intensität von den Drehzahlen 129

6.2.4 Variationsuntersuchungen der Makroinstabilitäten *131*

    6.2.4.1   Einführung 131

    6.2.4.2   Ergebnisse und Diskussion 131

6.3 WANDSTRAHLENSTRÖMUNG STRÖMUNG VOR DEN STROMBRECHERN 141

6.3.1 Einführung *141*

6.3.2 Ergebnisse zum gemittelten Geschwindigkeitsprofil *142*

6.3.3 Ergebnisse zum halb-sekundären Geschwindigkeitsprofil *145*

6.4 MÖGLICHKEITEN ZUR VERMEIDUNG DES GASEINTRAGS 147

6.4.1 Wellenstrombrecher *147*

6.4.2 Oberflächengitter *149*

7   ZUSAMMENFASSUNG **153**

8   LITERATUR **157**

ANHANG **171**

ANHANG A: RUNGE-KUTTA METHODE MIT NEWTON-RAPHSON SUCHFUNKTION 171

ANHANG B: LÖSUNG DER ODE MIT FV-VERFAHREN MIT TDMA 176

ANHANG C: LOMB-SPEKTRAL-ALGORITHMUS 182

SUMMARY IN GERMAN **185**

INHALTVERZEICHNIS 187

1   EINLEITUNG 191

1.1   Allgemeine Einführung *194*

1.3   Inhalt der Dissertation *192*

7   ZUSAMMENFASSUNG 197

1 EINLEITUNG

1.1 Allgemeine Einführung

Rührwerke spielen als essentielle Bestandteile der Anlagentechnik in nahezu allen Bereichen der chemischen Industrie, der Nahrungsmittelindustrie, der Biotechnologie und der Umwelttechnik eine wichtige Rolle. Trotz der Vielzahl äußerst unterschiedlicher Einsatzmöglichkeiten von Rührsystemen für die verschiedensten Anwendungsfelder, lassen sich nach Smith (1999) [117] und Zlokarnik (1999) [152] nachfolgende fünf Rühraufgaben klassifizieren:

- Homogenisieren, Ausgleich von Konzentrations- und Temperaturunterschieden mit oder ohne chemische Reaktionen,

- Suspendieren und/oder Aufwirbeln von Feststoffen in einer Flüssigkeit,

- Dispergieren eines Gases in einer Flüssigkeit in der Form von kleinen Blasen,

- Dispergieren von zwei nicht miteinander mischbaren Flüssigkeiten, z. B. Emulgieren,

- Intensivierung des Wärmeaustauschs.

Beim Homogenisieren und Emulgieren sind nur eine oder mehrere flüssige Phasen an den Prozessen beteiligt. Bei den anderen Rühraufgaben treten in der Regel zusätzlich noch Gase und/oder Feststoffe auf. Ein Rührsystem muss mindestens eine der obengenannten Rühraufgaben erfüllen. In der Praxis müssen häufig mehrerer Rühraufgaben gleichzeitig erfüllt werden, die jeweils von unterschiedlicher Bedeutung für den Gesamtprozess sind. Die zu bewältigenden Rühraufgaben sind die Grundlage für die Auswahl und Auslegung von Rührapparaten und den einzustellenden Betriebsbedingungen. Aufgrund der Komplexität der Zusammenhänge zwischen den verschiedenen Rühraufgaben und den unterschiedlich orientierten Auslegungsstrategien, gestaltet sich eine Auswahl und Optimierung von Rührapparaten oft als unmöglich, bei der alle geforderten Prozessaufgaben gleichzeitig erfüllt werden [125]. Eine passende Strategie der Optimierung und Auslegung, die sich an den wichtigsten Aufgaben orientiert, erfordert detaillierte Information der Strömungsfelder in gerührten Behältern.

Aufgrund der obengenannten Bedeutung gerührter Strömungen wurden seit jeher zahlreiche Untersuchungen zu den Strömungsfeldern in Rührbehältern durchgeführt, um die erforderlichen Detailinformationen über die Strömungsbedingungen zu erhalten. Diese Untersuchungen wurden in den letzten Jahren durch die jüngsten Entwicklungen von LDA-, PDA- und PIV-Messtechniken vorangetrieben, mit denen die in der Vergangenheit nicht vorhandenen experimentellen Methoden bereitgestellt werden konnten [26, 97-111, 120, 144-146, 98, 99, 133, 147, 87]. Des weiteren sind numerische Verfahren in die Strömungsuntersuchungen eingeführt worden und auch für die Berechnung gerührter Strömungen eingesetzt worden. Mit Hilfe der numerischen Strömungssimulationen in gerührten Behältern wurden wichtige Erkenntnisse über die wesentlichen Strömungsphänomene in Rührbehältern gewonnen und ein tieferes Verständnis über Rührerströmungen erreicht [4, 6, 24, 114, 133-136]. Die wesentlichen

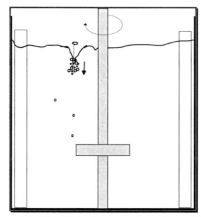

Abbildung 1.1: Oberflächenbegasung in gerührten Behältern

Charakteristika von einphasigen Rührerströmungen wurden auf der Grundlage experimenteller und numerischer Methoden durch unterschiedliche Forschungsgruppen erfolgreich untersucht, woraus zahlreiche rührtechnische Verbesserungen resultierten [144]. Allerdings blieben sowohl in den experimentellen als auch in den numerischen Untersuchungen bis heute viele instationäre Phänomene von Rührerströmungen unklar. Dazu gehört unter anderem der Gaseintrag über die freie Oberfläche, die niederfrequenten Strömungsoszillationen [12] und das sogenannte „Precessing-Vortex-Phänomen" [117].

Es ist allgemein bekannt, dass Umgebungsluft in offenen oder Gas innerhalb von geschlossenen Rührbehältern in die Flüssigkeit über die freie Oberfläche eingetragen wird, wenn die Rührerdrehzahl einen bestimmten Grenzwert übersteigt. In Abbildung 1.1 ist dieser Vorgang graphisch verdeutlicht, der bereits von Calderbank (1958) [17] als Oberflächenbegasung bezeichnet wurde. Oberflächenbegasung tritt häufig schon bei einer Rührerdrehzahl auf, die zur Erzielung einer ausreichenden Durchmischung des Rührgutes oder zur Erfüllung bestimmter anderer Kriterien, wie z.B. das Suspendieren eines Feststoffs, erforderlich ist. Daher kann die Oberflächenbegasung bei bestimmten Betriebsbedingungen die Qualität des Rührprozesses wesentlich beeinflussen.

Das Fehlen zuverlässiger Informationen über die Oberflächenbegasung war die Motivation zu dieser Arbeit. Im nachfolgenden Abschnitt wird zunächst eine Zusammenfassung einer zu Anfang der Arbeit durchgeführten Literaturstudie erfolgen. Die vorliegende Arbeit war von großer Bedeutung für ein Projekt aus der Beschichtungsindustrie, das am LSTM-Erlangen durchgeführt wurde [50]. Die zentrale Aufgabenstellung in diesem Projekt war die Vermeidung unerwünschter Oberflächenbegasung beim Anmischen der Beschichtungsrezeptur in einem Rührbehälter. Die Literaturstudie zu diesem Thema zeigte, dass trotz der großen Forschungsbemühungen nur wenige umfassende Untersuchungen zum Mechanismus des Gaseintrags in gerührten Behältern hinsichtlich der Strömungsmechanik vorlagen. Der erste, die Oberflächenbegasung betreffende Artikel erschien schon im Jahre 1958 [17]. Diese und nachfolgende Arbeiten beschränkten sich allerdings auf die vereinfachten Methoden der Dimensionsanalyse und die Ermittlung empirischer Korrelationen, die zudem auf rein visuellen Beobachtungen beruhten. Zahlreiche Korrelationen wurden erarbeitet, welche die Beziehungen zwischen der kritischen Rührdrehzahl und den wichtigsten geometrischen Verhältnissen, wie z.B. dem Rührertyp, dem Rührerdurchmesser und der Einbauhöhe des Rührelements, einerseits und den Betriebsbedingungen, wie z.B. den physikalischen Eigenschaften der beteiligten gerührten Phasen, andererseits darstellen. Die veröffentlichten Korrelationen sind jedoch von signifikanten Abweichungen bei unterschiedlichen Rührsystemen geprägt, welches im wesentlichen auf das Fehlen der physikalischen Grundlagen zur Oberflächenbegasung zurückzuführen war. Damit standen in der Vergangenheit nur beschränkte Auslegungs- und Optimierungsrichtlinien für die Praxis zur Verfügung.

Hieraus resultiert ein dringender Forschungsbedarf, um die Mechanismen des Gaseintrags in gerührten Behältern besser zu verstehen. Das Ziel der vorliegenden Arbeit ist es, grundlegende Auswahl- und Auslegungsrichtlinien für Rührsysteme bereit zu stellen, mit denen die Oberflächenbegasung unterdrückt und gleichzeitig eine ausreichende Durchmischung im Rührbehälter gewährleistet werden kann. Da bereits zahlreiche Korrelationsansätze vorliegen, mit denen sich der Einbruch des Gaseintrags vorhersagen lässt, wurde der Schwerpunkt der Arbeit darauf gelegt, die existierenden Korrelationen zu beurteilen und diese auf der Grundlage eines verbesserten physikalischen Verständnisses über die strömungsmechanischen Mechanismen zu modifizieren.

Der Autor ist davon überzeugt, dass die gewonnenen Ergebnisse nicht allein für das am LSTM-Erlangen durchgeführte Projekt von großem Interesse sind, da die Oberflächenbegasung sowohl positive als auch negative Auswirkungen besitzen kann. Mit den in dieser Arbeit vorgelegten Erkenntnissen können für diese Prozesse neue verbesserte Auslegungs- und Optimierungsrichtlinien zur Verfügung gestellt werden. Eine positive Eigenschaft der Oberflächenbegasung liegt darin, dass ein Gas-Flüssigkeits-Kontakt durch das Einbringen von Blasen aus der Umgebung bzw. dem Kopfraum des Behälters ermöglicht und/oder verstärkt wird. Dadurch lässt sich der Rührbehälterinhalt auch ohne irgend ein zusätzliches Belüftungssystem begasen. Diese Art von Begasungssystemen werden nach Forrester *et al.* (1997) [46] als Oberflächenbegasungssysteme bezeichnet. Oberflächenbegasungssysteme werden häufig in Bioabwasserbehandelungsanlagen und zur Durchführung chemischer Reaktionen eingesetzt, bei denen geringe Stoffübertragungsraten ausreichend sind [46], wie z.B. bei Hydrogenierungen, Chlorierungen und Oxidationen. Darüber hinaus ist es in den zuletzt genannten Fällen häufig anzustreben, das nicht umgesetzte Gas vom Behälterkopfraum in den Reaktor zurückzuführen, da das Gas z.B. hoch toxisch ist und dessen Freilassung zu Umweltproblemen führt, oder hohe Herstellungskosten aufweist und es sich damit lohnt, das Gas im Kreislauf zu fahren und vollständig umzusetzen [150]. Diese Beispiele zeigen, das die Charakteristik und Leistungsfähigkeit des Gaseintrags von besonderer Bedeutung sind und die Auslegung der Rührsysteme sollte sich für solche Prozesse hieran orientieren. Eine weiterführende Zusammenfassung findet sich bei Forrester *et al.* [46].

Neben den vorangehend beschriebenen positiven Effekten kann der Gaseintrag über die freie Oberfläche auch unerwünscht sein, wenn dadurch die charakteristischen Eigenschaften der ursprünglich im Behälter vorliegenden Flüssigkeit geändert werden. Als klassisches Beispiel lässt sich hierfür das Rühren in der Beschichtungsindustrie anführen. Schon das Vorhandensein sehr geringer Gasmengen während des Anmischens der Beschichtungsflüssigkeit führt zu einer signifikanten Beeinträchtigung des Beschichtungsprozesses und die geforderte Produktqualität kann dadurch nicht mehr gewährleistet werden. In diesen Fällen sind in der Regel zusätzliche Behandlungsschritte zur Entgasung der Beschichtungsflüssigkeit erforderlich, die aufgrund der hohen Viskositäten von Beschichtungsflüssigkeiten sehr aufwendig sind. Bei stärkeren Gaseinträgen gelangen die Gasblasen bis in das Gebiet des Rührelements, welches in einer deutlich verringerten Leistungsaufnahme des Rührers resultiert. Dieses kann immer dann zu Problemen führen, wenn gleichzeitig andere Rühraufgaben erfüllt werden müssen, in denen eine ausreichende Leistungsaufnahme gewährleistet werden muss, wie z.B. beim Suspendieren eines Feststoffs oder beim Wärmeaustausch. Darüber hinaus kann es durch ausgeprägte Blasenbildung hinter den Rührblättern zu Kavitationen kommen, die mechanische Probleme wie Wellenschwingungen verursachen.

1.3   Inhalt der Dissertation

Da es sich bei der Rührerströmung allgemein um eine rotierende Strömung handelt, werden im nachfolgenden Kapitel zunächst die Grundlagen und Eigenschaften von Rotationsströmungen eingeführt. Als eine gemeinsame Eigenschaft rotierender Strömungen gelten Instabilitäten, die insbesondere im Zusammenhang mit Ergebnissen neuerer Untersuchungen für rotierende Scheibenströmungen behandelt werden. Die durch die Instabilitäten hervorgerufene Wirbelablösung und das „precessing-motion" werden dabei vorgestellt. Darüber hinaus wird die Turbulenz und die Charakterisierung der turbulenten Rührströmungen sowie die Auswirkung der Turbulenz auf den Gaseintrag in diesem Kapital zusammengefasst. In Kapitel 3 werden die Grundlagen und die Anwendungen experimenteller Verfahren dargestellt, die in der vorliegenden Arbeit eingesetzt wurden. Des weiteren werden die in der vorliegenden Arbeit verwendeten numerische Verfahren erläutert, welches die Grundlagen des eingesetzten Finite-Volumen-Verfahrens und die Details der Gittererzeugung umfasst.

Im Kapitel 4 werden die einführenden Untersuchungen zu Rührerströmungen behandelt. Die Betrachtung einer durch eine rotierende Scheibe erzeugten Strömung, die als eine stark vereinfachte Strömung gilt, ermöglicht es, grundlegende Informationen über die Instabilitäten von Strömungsfeldern in gerührten Systemen zu erhalten. Dabei werden sowohl analytische als auch numerische und experimentelle Methoden eingesetzt. Die „precessing-motion" des Wirbelzentrums wird sowohl in den experimentellen als auch in den numerischen Untersuchungen beobachtet. Im fünften Kapitel werden die Beobachtungen in gerührten Behältern mit Standardrührsystemen dargestellt. Der Mechanismus der Oberflächenbegasung wird durch visuelle Beobachtungen studiert und durch analytische und experimentelle Untersuchungen bestätigt. Es wird zusammengefasst, dass die Anfangsoberflächenbegasung mit der „precessing-motion" des Wirbelzentrums zu verbinden ist. Basierend auf den Ergebnissen der visuellen Beobachtungen wird ein aus der Dimensionsanalyse abgeleiteter, physikalischer Ansatz vorgestellt und überprüft.

Im Kapitel 6 werden Rührerströmungen hinsichtlich der Bildung von Oberflächenwirbeln behandelt. Dieses beinhaltet eine Analyse und Charakterisierung der Strömungsfelder an der Oberfläche, der Makroinstabilitäten in gerührten Behältern als eine Darstellung des „precessing vortex core", der Intensität der Makroinstabilitäten sowie der Wandstrahlströmungen vor den Strombrechern, welche zur Bildung der Oberflächenwirbel erheblich beitragen. Darüber hinaus werden zwei Möglichkeiten hinsichtlich der Vermeidung der Oberflächenbegasung in gerührten Behältern abgeleitet und überprüft. Kapitel 7 fasst die wichtigsten Ergebnisse der vorliegende Arbeit zusammen. Abschließend werden in Kapitel 7 Empfehlungen für weiterführende zukünftige Arbeiten angedeutet.

7 ZUSAMMENFASSUNG

Die vorliegende Arbeit behandelte zunächst Untersuchungen zur Vermeidung der Oberflächenbegasung in gerührten Behältern. Auf der Grundlage der einführenden Untersuchungen konnte der Mechanismus der Oberflächenbegasung aufgeklärt und erstmalig mit dem „precessing vortex core" (PVC) und den Makroinstabilitäten in gerührten Behältern in Verbindung gebracht werden. Die wichtigsten Ergebnisse der vorliegenden Arbeit lassen sich wie folgt zusammenfassen:

Mechanismus der Oberflächenbegasung und Makroinstabilitäten in Rührbehältern

- Unterschiedliche Formen von Instabilitäten treten in rotierenden Strömungen auf. Eine axialsymmetrische Strömung tendiert dazu, ihre Stabilität zu verlieren und trotz der Axialsymmetrie in den Randbedingungen unsymmetrisch zu werden. Das PVC, das üblicherweise mit der Wirbelablösung verbunden ist, ist eine Folge der Instabilität in wirbelbehafteten Strömungen. In der vorliegenden Arbeit wurde die rotierende Scheibenströmung in einem rechteckigen Behälter theoretisch, numerisch und auch experimentell untersucht. In den numerischen Untersuchungen zeigte sich, dass die Strömung zu Instabilitäten neigt, wenn die Drehzahl der Scheibe über einen bestimmten Grenzwert steigt und ohne dass irgend eine zusätzliche Störung aufgebracht werden muss. Das Rotationszentrum der Strömung wird dabei in eine „precessing motion" versetzt. Das PVC-Verhalten wurde durch experimentelle Untersuchungen sowohl für die rotierende Scheibenströmung als auch für Strömungen in bewehrten Rührbehältern nachgewiesen.

- Der Gaseintrag wird hauptsächlich durch das Zusammenwirken von Oberflächenwirbeln und Störungen in der Rührbehälterströmung verursacht. Bei radialfördernden Rührelementen erzeugt das PVC einen dominanten Oberflächenwirbel, an dessen Fuß Blasen aufgrund der Störungen in der Hauptströmung in die Flüssigkeit eingebracht werden. Für axialfördernde Rührelemente ist die Rotation der Strömung nicht derart ausgeprägt wie bei radialfördernden Rührelementen, so dass an der Oberfläche eine dominante rein radial ausgeprägte Strömung ohne Rotation vorherrscht, deren Richtung an der Oberfläche unregelmäßig über der Zeit variiert. Durch diese Strömung entstehen Nachlaufwirbel hinter den Strombrechern. In beiden Fällen kann der Gaseintrag auf das Vorhandensein von Makroinstabilitäten zurückgeführt werden.

- Die Oberflächenbegasung in gerührten Behältern lässt sich durch fünf Teilvorgänge beschreiben, in denen die nachfolgenden Mechanismen und Voraussetzungen eine entscheidende Rolle für die einzelnen Teilvorgänge spielen:

- ◆ Änderung der Oberflächenform zur Ausbildung von Oberflächenwirbeln,

♦ ein ausreichender Turbulenzgrad für den Blaseneintrag über die Oberflächenwirbel,

♦ ausreichende Kräfte zur Überwindung der Oberflächenspannung zum Aufbrechen der Blasen und

♦ ausreichende abwärts gerichtete Strömungsgeschwindigkeiten zum Transport der einge-tragenen Blasen in das Rührergebiet

- Ein physikalischer Ansatz, in dem alle obengenannte Effekte bei der Oberflächenbe-gasung berücksichtigt werden, wurde in der vorliegenden Arbeit vorgestellt und über-prüft. Diese Korrelation lautet:

$$Fr = CWe'^{\alpha}Mo^{\beta}\left(\frac{H-C}{D}\right)^{\gamma}\left(\frac{T}{D}\right)^{\delta}.$$

Die Koeffizienten sind für die jeweils relevanten geometrischen Verhältnisse und Be-triebsbedingungen durch Experimente zu ermitteln.

Makroinstabilitäten in gerührten Behältern

- Die Zeitreihen der Geschwindigkeiten enthalten niederfrequente Schwankungen, de-ren Zeit- und Längenmaße diejenigen der Turbulenz signifikant überschreiten. Diese Art von niederfrequenten Variationen werden durch das PVC hervorgerufen und erstrecken sich über den ganzen Rührbehälter. Das gesamte Strömungsfeld, das auf-grund der Rotation des Rührelements periodische Schwankungen aufweist, wird durch die niederfrequenten Schwankungen überlagert.

- Die niederfrequenten Schwankungen lassen sich durch Spektralanalyse charakterisie-ren. Bei radialfördernden Rührelementen ist eine dominante Schwankungsfrequenz vorhanden, die der Rotationsfrequenz der dominanten Oberflächenwirbel entspricht. Dadurch lässt sich die inhärente Verknüpfung zwischen dem PVC-Phänomen und der MI in gerührten Behältern nachweisen. Die konstante modifizierte Strouhalzahl f_{MI}^{*} deutet auf die Unabhängigkeit der Strömung von der Reynoldszahl und den physikali-schen Eigenschaften des Rührgutes in geometrisch ähnlichen Konfigurationen hin. Bei axialfördernden Rührelementen sind die dominanten MI-Frequenzen von der Messdauer abhängig und weisen eine signifikante Streuung auf. Hieraus kann geschlossen werden, dass sich die dominante MI-Frequenz bei axialfördernden Rührelementen aus unterschiedlichen Frequenzen mit unterschiedlichen Dominanzni-veaus zusammensetzt.

- Eine Abschätzung der Intensität der niederfrequenten Schwankungen gelingt über eine Aufteilung der Geschwindigkeiten in einen Anteil, der aus den niederfrequenten Makroinstabilitäten resultiert, und einen Anteil, der durch die überlagerten turbulenten Fluktuationen vorhanden ist. Es wurde gezeigt, dass die Intensitäten der MI ein einheitliches Niveau im ganzen Rührbehälter aufweisen. Dagegen nimmt der Beitrag der rein turbulenten Schwankungen am lokal gebildeten RMS-Wert und an der turbulenten kinetischen Energie mit zunehmendem Abstand von dem Rührelement bzw. der Turbulenzquelle ab. Diese Erkenntnis ist von großer Bedeutung, wenn man experimentelle und numerischen Untersuchungen in Rührbehälterströmungen vergleicht, weil im letzten Fall der MI-Beitrag zur turbulenten kinetischen Energie offensichtlich nicht berücksichtigt und diese damit unterschätzt wird. Der Anteil der MI an der turbulenten kinetischen Energie kann in Abhängigkeit von der Position im Rührbehälter bis zu 80% betragen. Das weist darauf hin, dass die Makroinstabilität der Hauptmechanismus des Mischens im oberen Bereich des Rührbehälters ist.

• Die Geometrienbedingungen haben einen signifikanten Einfluss auf die dominante Frequenz der Makroinstabilitäten. Insbesondere spielt dabei die Einbauhöhe des Rührelements eine übergeordnete Rolle. Die dominante Frequenz springt zwischen unterschiedlichen Stufen. Bei symmetrischen Einbaubedingungen für das Rührelement erreicht die Resonanzfrequenz ihr höchstes dominantes Niveau.

Wandstrahlströmungen vor den Strombrechern

• Die Wandstrahlströmungen vor den Strombrechern tragen erheblich zur Störung der Oberflächenwirbel und den hieraus resultierenden Gaseintrag über die Oberfläche bei. Die Wandstrahlströmungen lassen sich durch die Ähnlichkeitstheorie charakterisieren, mit der der Strömungszustand bestimmt werden kann. Beim Sechs-Blatt-Scheibenrührer bleiben die Wandstrahlströmungen bei einer Reynoldszahl von $Re = 5424$ nur bis zu etwa Zweidrittel des Rührbehälters voll turbulent. Beim Schrägblattrührer ist die volle Turbulenz bei $Re = 7322$ nur noch über die Hälfte des Rührbehälters existent. Unter Berücksichtigung der Makroinstabilitäten, die die zeitlichen Änderungen in den Wandstrahlströmungen verstärken, werden die Strahlströmungen an der Oberfläche turbulenter, so dass sie die Oberflächenwirbel in Form aufwärts gerichteter Stoßbewegungen gestört werden.

Unterdrückung der Oberflächenbegasung

• Aus den gewonnenen Ergebnissen konnten zwei Strategien zur Vermeidung der Oberflächenbegasung abgeleitet werden:

(1) die Unterdrückung der Ausbildung von Oberflächenwirbeln und

(2) die Absenkung des Turbulenzgrades in der Nähe der Oberflächenwirbel

- Entsprechend dieser zwei Strategien wurden in der vorliegenden Arbeit zwei praktische Lösungsansätze für die Vermeidung von Oberflächenbegasung zum Einsatz gebracht:

    (1) Wellennahe Strombrecher und

    (2) Oberflächengitter.

    Durch den Einsatz von Oberflächengittern werden die rein turbulenten Schwankungen in der Nähe der Oberflächenwirbel unterdrückt. Diese Anwendung ist daher für alle Rührelementtypen einsetzbar. Dagegen werden bei wellennahen Strombrechern beide Strategien verfolgt. Aufgrund der Strömungsverhältnisse wirken diese allerdings nur bei radialfördernden Rührelementen.

Empfehlungen für zukünftige weiterführende Arbeiten

- Makroinstabilitäten sind auch für weitere instationäre Vorgänge in gerührten Behältern verantwortlich, wie z.B. die enormen Schwankungen von Makromischzeiten, instabile Vorgänge am Behälterboden beim Suspendieren von Feststoffen und Schwingungen an der Rührerwelle. Der detaillierten Charakterisierung von Makroinstabilitäten und ihrer Intensitäten kommt daher eine große Bedeutung zu, um zu einem besseren Verständnis über die instabilen Vorgänge in gerührten Behältern zu gelangen. Dies sollte zunächst in vereinfachten Geometrien durchgeführt werden, wie z.B. in Wirbelkammern, für die Alekseenko *et al.* [2] sogar in der Lage waren, die dominante Frequenz analytisch vorherzusagen. Die Untersuchungen können durch den Einsatz numerischer Verfahren unterstützt werden, insbesondere durch Large-Eddy-Simulationen (LES). Sobald sich die Makroinstabilitäten modellieren lassen, können die entwickelten Modelle in die Standard-CFD-Programme für Rührerströmungen integriert werden, um die Leistungsfähigkeit der Berechnungen in der Rühr- und Mischtechnik zu verbessern.

CURRICULUM VITAE

Jian Yu

geb. am 17.12.1970 in Dandong, Liaoning, VR China
Staatsangehörigkeit: Chinesisch
Familienstand: verheiratet, ein Kind

Berufstätigkeit

04/1995 – 08/1996	Mitarbeiter der Firma LIAONING FOREIGN TRADE INVESTMENT CO. in Dalian, Liaoning, VR China
seit 05/1998	Wissenschaftlicher Mitarbeiter am Lehrstuhl für Strömungsmechanik in Erlangen

Ausbildung

09/1996 – 07/1997	Besuch eines Intensivkurses des Deutschen Sprachzentrums an der Technischen Universität Beijing
	Abschluss: DSH (Deutsche Sprachprüfung für den Hochschulzugang ausländischer Studienbewerber)
08/1997 – 04/1998	Vorbereitung auf die Bewerbung für ein Stipendium beim DAAD (Deutscher Akademischer Austauschdienst) in Beijing und Shanghai

Hochschulausbildung

09/1988 – 07/1992	Studium an der Technischen Universität Dalian, VR China
	Studienrichtung: Maschinenbau
	Abschluss: B.Sc. (Eng.) in Mechanical Engineering
09/1992 – 03/1995	Studium an der Technischen Universität Dalian, VR China
	Studienrichtung: Maschinenbau
	Abschluss: M.Sc. (Eng.) in Mechanical Engineering

Schulausbildung

1977 – 1982	Jintang Grundschule in Dandong, VR China
1982 – 1984	Mittelschule Nr. 13 in Dandong, VR China
1984 – 1988	Mittelschule Nr. 2 in Weihai, VR China

Stuttgart, den 18. Januar 2003